CW01572610

Test Equipment for the Radio Amateur

Fifth Edition

by *Philip Lawson* G4FCL

Radio Society of Great Britain

Published by the Radio Society of Great Britain, 3 Abbey Court, Fraser Road, Priory Business Park, Bedford MK44 3WH. Tel 01234 832700 *www.rsgb.org*

First Edition Published 1974

This edition published 2018

Editing: Paul Lee, G4EJB

Layout: Mike Browne, G3DIH

Cover Design: Kevin Williams, M6CYB

Production: Mark Allgar, M1MPA

Printed in Great Britain by Latimer Trend Ltd of Plymouth

ISBN: 9781 9101 9365 5

Contents

Preface

Recent editions of this book were developed into a highly detailed and authoritative text thanks to the considerable work of Clive Smith, GM4FZH. Going forward, the new author has taken the opportunity to recast the book, separating the description of commercial test equipment and how it is used, from home-brew test equipment construction projects. This is to aid clarity and so that these two topics can be developed further.

In this edition, increased emphasis is placed on practical measurement technique; or in other words, what you can do with the equipment, how to do it, and things to watch out for. A few items of test equipment and measurement technique that have become dated and fallen out of use have been removed, while those which are likely to form the backbone of average Radio Amateur's shack or workshop have obviously been retained and revisited.

While a lucky few may have access to advanced and expensive test equipment (and know how to use it!), most will not, and so considerable emphasis is placed on making the best use of equipment that is available commercially at reasonable cost. Happily, the price of some hitherto unaffordable equipment continues to fall, such that an old surplus model, or a new solid-state version, is now a viable purchase for many. While the latest widget may help you to make measurements faster and more easily, older equipment available at a fraction of the price is often perfectly satisfactory - provided you have the knowledge and skill to use the equipment properly and are aware of its limitations. This book can help you with that.

Self-build, either from scratch or from a kit is not neglected, as this is great way to increase your capability at low cost, as well as being a wonderful learning experience. Many time-less constructional projects from previous editions have been imported into this edition, but as the internet has become an integral part of reader's lives, the opportunity has been taken to link-out to sources of original articles (and their support pages), rather than reproduce them in full here. This enables the reader to access the very latest information and help on a particular project.

The appendices contain a wealth of reference information, including component identification data and some theory to help in the understanding of certain measurements. PCB and component layout schematics are also provided for a few projects.

Whether you are just starting out, or an old hand, I very much hope that you will find this new edition of *Test Equipment for Radio Amateurs* both interesting and useful.

Philip Lawson, G4FCL
November 2018

Acknowledgements

Where an article has been referenced or reproduced, in whole or in part, the author has endeavoured to credit the source (where known) in the appropriate section of this book. Particular thanks go to those who contributed to previous editions, some material from which is carried over into this edition, and to the authors of more recently published *RadCom* articles which provided the very latest source material. The author is also grateful to Clive Smith GM4FZH for his suggestions at handover, and to Lee Aldridge, G4EJB for his extremely helpful review.

About the Author

Influenced by his father, who was a keen repairer of Black & White TV sets in his spare time, Philip naturally developed an interest in electronics at an early age. Licensed in the early 1970s, initially as G8ING and later as G4FCL, he was a keen member of Derby & District Amateur Radio Society, winning several trophies for Amateur station operating and radio construction. After university, Philip took up a professional career in electronic engineering and is now a Chartered Engineer with over 30 years' experience leading RF, Microwave and Optical product development in telecoms, aerospace, and defence. He has published over 40 internal and external reports and his achievements include several world firsts.

Phillip has another book published by the RSGB *Restoring Old Radio Sets* copies of which can be obtained from the RSGB shop.

Eur Ing Philip Lawson BSc(Eng) MSc CEng CPhys MIET MInstP PgDLaw LLM

Glossary - Abbreviations and Units

Abbreviations

AC	Alternating Current	IF	Intermediate Frequency
A/D	Analogue To Digital	IP	Intermodulation Product
ADC	Analogue To Digital Converter	LCD	Liquid Crystal Display
AF	Audio Frequency	LED	Light Emitting Diode
AFC	Automatic Frequency Control	LF	Low Frequency
AFSK	Audio Frequency Shift Keying	MOSFET	Metal Oxide Semiconductor FET
AGC	Automatic Gain Control	NF	Noise Figure
AM	Amplitude Modulation	PC	Personal Computer
ARRL	American Radio Relay League	PCB	Printed Circuit Board
CB	Citizen's Band	PEP	Peak Envelope Power
CMOS	Complementary Metal Oxide Semiconductor	PIV	Peak Inverse Voltage
		PPM	Parts Per Million
CRO	Cathode Ray Oscilloscope	PSD	Phase Sensitive Detector
CRT	Cathode Ray Tube	PSU	Power Supply Unit
CW	Carrier Wave	PTFE	Polytetrafluorethylene
DC	Direct Current	PTT	Press To Talk
DIL	Dual In Line	QRP	Low Power
DMM	Digital Multimeter	RAM	Random Access Memory
DSP	Digital Signal Processing	RCD	Residual Current Device
DVM	Digital Volt Meter	RF	Radio Frequency
ECW	Enamelled Copper Wire	RLB	Return Loss Bridge
EMC	Electromagnetic Compatibility	RMS	Root Mean Square
EMI	Electromagnetic Interference	RSGB	Radio Society of Great Britain
ENR	Excess Noise Ratio	SINAD	SIgnal and noise and distortion, to Noise And Distortion ratio
ERP	Effective Radiated Power		
ESR	Equivalent Series Resistance	SMD	Surface Mount Device
ETSI	European Telecommunications Standards Institute	SMPS	Switched Mode Power Supply
		SNR	Signal to Noise Ratio
FET	Field Effect Transistor	SSB	Single Side Band
FFT	Fast Fourier Transform	SWG	Standard Wire Gauge
FM	Frequency Modulation	THD	Total Harmonic Distortion
FSD	Full Scale Deflection	TTL	Transistor-Transistor Logic
GDO	Grid Dip Oscillator	VCO	Voltage Controlled Oscillator
GPS	Global Positioning System	VFO	Variable Frequency Oscillator
GPSDO	GPS Disciplined Oscillator	VHF	Very High Frequency (30-300MHz)
GND	Ground		
HF	High Frequency (3-30MHz)	VNA	Vector Network Analyser
HRC	High Rupture Capacity	VSWR	Voltage Standing Wave Ratio
HT	High Tension	UHF	Ultra-High Frequency (300-3000MHz)
IC	Integrated Circuit		

Electrical Units

A	Ampere	current
F	Farad	capacitance
H	Henry	inductance
Hz	Hertz	frequency
s	second	time
S	Siemen	admittance, conductance, susceptance
V	Volt	voltage
Ω	Ohm	resistance, reactance, impedance

Decimal Multipliers

T	Tera	10^{+12}
G	Giga	10^{+9}
M	Mega	10^{+6}
k	kilo	10^{+3}
m	milli	10^{-3}
μ	micro	10^{-6}
n	nano	10^{-9}
p	pico	10^{-12}

Pointers

- The arrow sign '→ ' is used to indicate where to find further information on a topic.

- Reference is occasionally made to excellent articles in the Radio Society of Great Britain's monthly journal *RadCom*. Members will have received a hard copy and have on-line access to the back-issues cited. Non-members can obtain an electronic copy of the yearly journal archive from RSGB Sales at: *https://www.rsgbshop.org*

1 Introduction

ASIDE FROM technical issues, great personal satisfaction can be gained from the skilful handling of test equipment. A user who understands the operation and limitations of his or her test equipment has the ability not only to make measurements, but more importantly, the ability to make the right measurements, and to correctly interpret their meaning.

As Radio Amateurs, we know that test equipment is essential for activities such as checking and demonstrating compliance with the licence conditions, occasional servicing of equipment, and testing new self-build projects; but we should not forget that Amateur Radio is above all a leisure activity [1], and that the more we understand, the more we are likely to enjoy the hobby.

1.1 Safe Testing

This matter has to be addressed first, as danger lurks ready to pounce. If you are using test equipment, then almost by definition, you must have something on the bench to test. It could be anything from a mains transceiver to a small battery-powered home-construction project. So what are the hazards? Answering this question, so that safeguards can be thought about and put in place can be problematic, because the Radio Amateur is primarily trained in radio communication, not safety. Unlike in the workplace, the Amateur's environment is not subject to strict rules on operator training, process control, and storage requirements, with regular lab inspections and corrective feedback to maintain a safe and effective work-space. This absence of third-party control gives the Amateur enormous freedom to do what he/she likes, but with this freedom does comes a responsibility, and an obligation, not to endanger him/herself, and especially others, such as visitors or family members.

Many of the risks noted in this chapter and the precautions that should be taken against them, are of course common-sense; however, there are some risks which may not be so obvious, particularly to younger or inexperienced operators.

Testing equipment in the Amateur workshop or shack presents many potential or real hazards depending on what has to be done.

These hazards include (but are not limited to):

• Receiving an electrical shock, which could prove fatal

• Being cut by sharp metal projections from a chassis or glass edges

• Being struck and injured by a heavy cabinet or chassis

• Being blinded by an exploding component (such as a capacitor)

• Getting accidentally burnt on a hot resistor or soldering iron

• Fire and fumes caused by overheating or shorting components

• Fire caused by accidental ignition of flammable cleaning materials

• Fire from spontaneous combustion of rags soaked with certain oils

• Incapacity arising from breathing vapours from volatile liquids

• Short or long term illness caused by breathing or absorbing chemicals

• Exposure to white asbestos (present in some older equipment) and potentially asbestosis

• Exposure to RF radiation

• Exposure to laser radiation

These hazards are summarised in **Fig 1.1**.

A common situation is where an individual knows of a generic risk, but has failed to re-

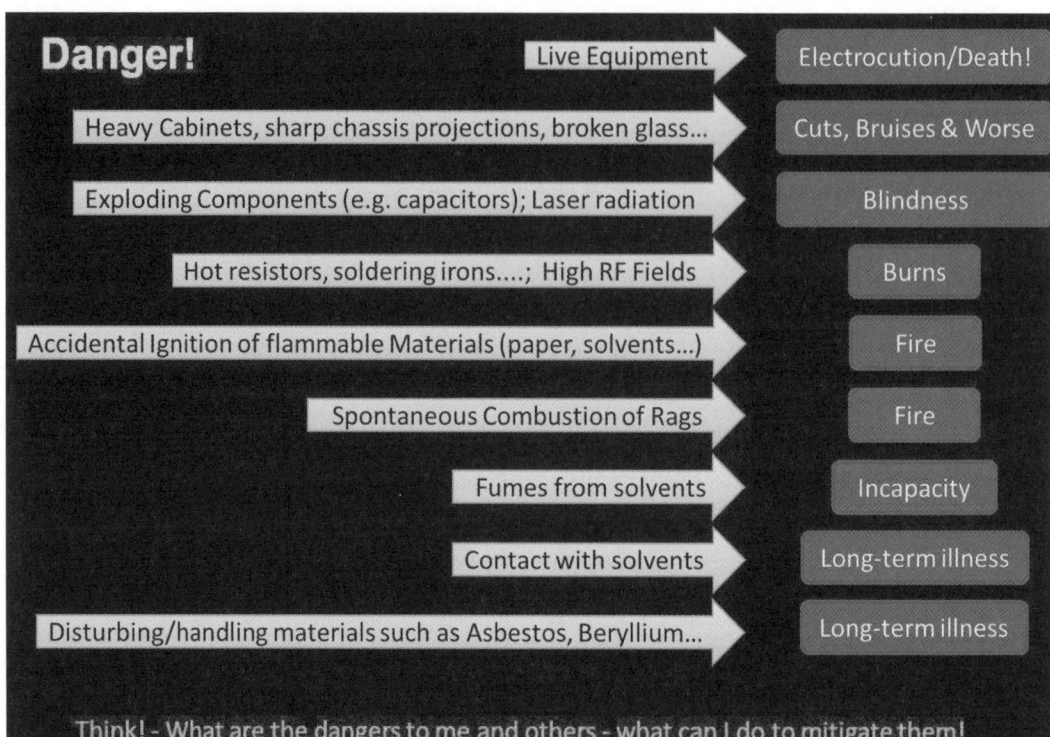

Danger!

Hazard	Consequence
Live Equipment	Electrocution/Death!
Heavy Cabinets, sharp chassis projections, broken glass...	Cuts, Bruises & Worse
Exploding Components (e.g. capacitors); Laser radiation	Blindness
Hot resistors, soldering irons....; High RF Fields	Burns
Accidental Ignition of flammable Materials (paper, solvents...)	Fire
Spontaneous Combustion of Rags	Fire
Fumes from solvents	Incapacity
Contact with solvents	Long-term illness
Disturbing/handling materials such as Asbestos, Beryllium...	Long-term illness

Think! - What are the dangers to me <u>and others</u> - what can I do to mitigate them!

Fig 1.1 Summary of testing hazards

alize, until the moment of disaster, that the risk applied to their situation. Therefore:

- <u>think ahead about what you intend to do</u>
- <u>review carefully all the risks in that situation and take steps to mitigate them</u>.

If you are in a family environment, you need to consider whether testing is actually a feasible proposition, and if you were to go-ahead, what additional measures you might need to put in place to keep your family safe. To help you deal with the issue of safety, here is some basic practical advice to start you off:

Practical Precautions

As we have noted, most readers will not have access to professional facilities that are strictly controlled by company rules and employee legislation. Instead, the reader will probably be working on something akin to the kitchen table or perhaps the luxury of a shed-workshop. So it would be wise to take the following precautions:

a) Organisation

It is very easy to cause accidental damage to yourself or to the equipment under test if the workspace is stuffy and cluttered. Therefore:

- Ensure that children, pets, and any unhelpful adults can be excluded for everyone's safety.

- Keep the working area tidy, obstacle-free, well ventilated, and well-lit.

- Do not have any more test equipment and materials out than is immediately necessary.

- Keep chemicals labelled and stored away from the working area (ideally in a special-purpose chemical-cabinet). If they are on the bench, only remove the top/stopper when needed to dispense the contents and firmly replaced at all other times.

- Think ahead before you touch anything.

- Don't work on live equipment if you're tired.

b) Electrical Shock

People have been electrocuted by contact with potentials as low as 35 Volts. The peak AC mains voltage in the UK is ten times that. The worse scenario is with AC current flowing through the heart from one hand to the other (the most likely scenario if you touch a live circuit). To make matters worse, if you grab the connection, the hand muscles tend to contract, increasing the current and making the effect far worse. So to protect yourself and others:

- Have a mains cut-off switch within immediate touching distance (tell others about it in case of emergency).
- If you don't already have RCD protection on your mains supply, fit RCDs to your mains sockets (these will trip above a certain leakage current).
- Use a neon mains-test screwdriver to check that the chassis, and exposed metal parts are not 'mains live'.
- Use a multimeter with well-insulated probes to check circuit voltage levels.
- Only touch live equipment with one hand at any time (to avoid current across the heart).
- Remember that capacitors store charge, and you can get a nasty shock long after the equipment is disconnected. Use a wand (an earthed probe) to discharge them.

c) Physical Injury

Don't under-estimate how easy it is to cause accidental damage to yourself and possibly others:

- Wear sturdy gloves to avoid being cut by a heavy cabinet, a sharp chassis, or glass edges.
- Think carefully about how you're going to manoeuvre that heavy cabinet and/or chassis before you pick it up! Enlist help to share the weight and make the job easier.
- Wear goggles to avoid being blinded by an exploding component (such as a capacitor).

- Only power the soldering iron when you actually want to use it – this is to avoid burns caused by accidental contact with the hot end, or the sudden ignition of volatile chemicals.
- Don't touch components such as high-wattage resistors and valves which can burn – use an infra-red thermometer to remotely check their temperature.
- If you wear a ring, put a light insulating glove on. This is to prevent a serious burn caused by a heavy current passing through it, for example when working on a low-voltage power supply.

d) Fire

It is unbelievably easy to accidently start a fire. Unless you can put it out in seconds, you and others are likely to be in a very serious predicament.

- Have a fire extinguisher to hand (one compatible with electrical and chemical hazards)
- Be on constant look out for overheating or shorting components. These can ignite solvents and nearby flammable materials like newspaper or cleaning rags.
- Keep tops/stoppers on all bottles – many chemicals will contain flammable liquids. These are easily ignited by spillage (or even vapour alone) and heat or spark.
- Keeps rags soaked with oils in a sealed metal tin to avoid spontaneous combustion. Lay them out flat in the garden later, to dry out.

e) Incapacity/Illness

Don't ignore how you feel, and be on the lookout for things which are not good for you:

- Ideally, let someone know if you're working alone, so they can they can regularly check on you and raise the alarm if you become incapacitated.
- Ensure there is plenty of ventilation (ie do not work in a small stuffy room) because breathing vapours from volatile liquids can cause drowsiness, nausea, and incapacity.
- Try to avoid breathing-in, or contact with,

chemicals, to minimise the risk of a reaction, or contracting a long term-effect such as cancer.

• Clearly label bottles of home-made mixtures with their contents and a warning 'Poison' in large red letters to avoid them being mistaken for foodstuff.

• If you come across white asbestos, for example around mains droppers, do not touch it and do not disturb it, to avoid exposure to the risk of asbestosis.

f) RF and Laser Radiation

The link between electromagnetic radiation in the environment (phone masts, smart phones, power lines etc), and its effect upon human physiology is not well understood, but the risk is generally considered quite small, but even so, it remains common-sense to minimise the level, and time, exposed to RF radiation. Therefore:

• Try to avoid working on antennas that are energised. There is danger both from high RF voltage and tissue damage. The same goes for power amplifiers (especially at VHF/UHF) with their covers removed.

• At UHF/microwave/mm-wave, never look into the open end of an activated waveguide, or antenna, or point them at anyone.

• Ensure that dummy loads are well shielded.

As a guide, between 10 and 146MHz, avoid exposure to field strengths >20V/m, which means keeping at least 5m away from an antenna radiating 200W ERP.

Lasers are potentially dangerous because their beam is usually concentrated into a diameter less than that of the pupil of the human eye, enabling it to cause damage to the retina. While the blink response of the human eye *might* provide protection to a visible laser (if lucky), there is no defence to a UV or other invisible laser. For these reasons, lasers should:

• Never be left on and unattended

• If visible, be run in expanded mode if possible (diverge to at least 50mm for safety)

• If invisible, only be run in expanded mode (for safety).

When working with lasers that are not explicitly eye-safe, safety goggles should be worn. The danger is not simply from direct viewing, but also from 'glint' – catching the reflection of the laser from some object in its path.

→ Search 'laser safety glasses/goggles' to find suitable suppliers.

g) Thunderstorms

Most Radio Amateurs will have an outdoor antenna, which is connected directly to their precious (and often expensive!) communications equipment. Ideally, some form of static discharge should be provided, or alternatively, the antenna should be unplugged *well before* the storm arrives - you really don't want to be holding the aerial plug at the time of a direct hit! The antenna should not be grounded, as this turns it into a lightning conductor, and is an invitation to a passing cloud to discharge. An additional precaution might be to unplug all sensitive/valuable equipment in case of a mains voltage surge.

→ For an account of what happens from someone who knows, see:

'Lightning Strikes at M0MCX', *RadCom*, April, 2018, p52.

h) In the Field Testing

When planning to operate 'away from base' for example, at a club rally, special event station, or DX-expedition, a full risk assessment should be undertaken beforehand and appropriate insurance cover obtained.

Remember that risks obvious to the Radio Amateur will not necessarily be obvious to others. A primary concern must be to manage the safety of members of the public, especially children, and the elderly. Failure to do so could result in accidents, and expose those involved in the organisation of the event to legal liability. Things to ask might include:

• Are antennas (and their supporting poles and guy-wires) sited well away from overhead lines *and* from members of the public?

Are they secure against strong winds and will running water from heavy rain pose a problem or safety threat?

• Are there any exposed electrical conductors that little fingers might touch? Can the conductors be shielded in some way?

• Is the equipment always under qualified supervision, to prevent young, and perhaps older interlopers, from 'having a go' when someone's not looking?

• Are power cables kept off the ground? Where they must run on the ground, are they covered with a cable protector to minimize the trip hazard?

• For outside installations where the ground may be wet, is an RCD with fast disconnect fitted? Are all mains connectors undamaged and (if necessary) waterproofed? Are the fuse ratings as low as possible (eg 5A rather than 13A) for better protection.

• And so on.

The list of what could go wrong, and the associated list of what should be done to prevent the risk from being realised, should be considered very carefully indeed for everyone's benefit. The list should be recorded and kept safe as evidence of responsible organisation in the event of an accident, investigation, and claim. In safety terms, an uneventful event is a good event, and is the reward for acting on a good risk assessment.

Some Sources of Help

The RSGB, the government, and the Health & Safety Executive (amongst others) have useful information packs to help assess the risk and keep you and others safe. See:

→ Safety in the shack (RSGB):

http://rsgb.org/main/get-started-in-amateur-radio/setting-up-your-shack/safety-in-the-shack/

→ Earthing (RSGB)

http://www.rsgb.org/main/files/2012/11/UK-Earthing-Systems-And-RF-Earthing_Rev1.3a-.pdf

http://www.rsgb.org/main/files/2012/11/EMC07-v3-Earthing-and-the-Radio-Amateur-Basic.pdf

→ Antenna safety (RSGB):

http://rsgb.org/main/get-started-in-amateur-radio/antennas/antenna-safety/

→ For a sobering look at what you could be getting into from a legal perspective, see:

'Elf 'n' Safety and Putting Up Aerials', Steve Hartley, G0FUW, *RadCom*, July 2016.

→ Laser radiation safety (UK government):

https://www.gov.uk/government/publications/laser-radiation-safety-advice/laser-radiation-safety-advice

→ Club events – risk assessment (RSGB):

http://rsgb.org/main/clubs/events-pack/risk-assessment/

→ Safely managing an event (HSE):

http://www.hse.gov.uk/event-safety/managing-an-event.htm

1.2 Test Bench Tools, Facilities, and other Desirables

Having considered the risks, let's have a look at some of the tools, facilities, etcetera, that you might need to do the job. But what job? You could be checking the 'vital signs' on that PCB you've just built; fault-finding a piece of kit; conducting a transmitter/receiver alignment and calibration exercise; making system measurements; or perhaps repairing and restoring an old piece of equipment. **Fig 1.2** gives an overview of what you might need to do the job.

There's no definitive list to cover the infinite range of tasks that you may be faced with, but the more you have from the following, the quicker and easier the task will be:

Facilities

Try to have plenty of working space and keep those bits that you tend to use most close at hand. You are likely to need:

• A sturdy non-conductive table or workbench.

• A non-slip surface to walk around the work area

• Good bright lighting, from overhead fluorescent lights and/or an angle-poise lamp. If you use the latter, it *must* be LED (not incan-

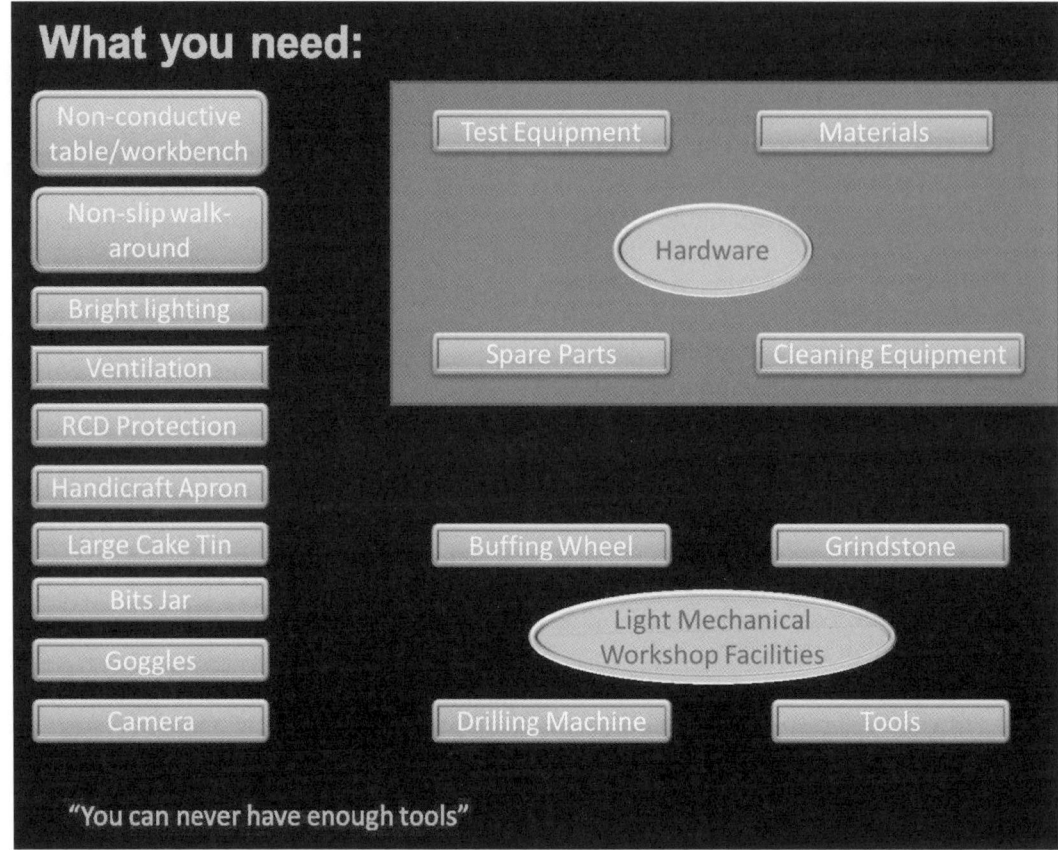

Fig 1.2 An overview of useful items

descent), or your hand and wrists will rapidly suffer from painful heat-stroke.

• A low-voltage spotlight which provides an intense beam over a small area, and mounted a couple of feet away, can be especially useful for the careful examination of circuits, and for tackling intricate soldering jobs.

• Plenty of ventilation for when using solvents.

• Mains supply with RCD protection.

• A handicraft apron, as it not only protects one's clothes while rubbing or cleaning bits, but it has useful pockets for tools to save your legs.

• A large jar and small containers to temporarily keep bits in that you remove (screws, washers, knobs, that sort of thing). Freezer bags - especially clear and re-sealable types - can be particularly useful.

• A large cake tin or similar for rags damp with solvent. Be aware that some oils have the potential to spontaneously combust, so keep rags which have been in contact with such liquids in a tin with the lid sealed down when working to prevent ignition. This practice also minimizes the health hazard from the volatile vapours that are given-off. You can dry them later by fully outstretching them and laying them flat, outside in the garden.

• Ideally, access to a buffing wheel (for polishing knobs etc) and a grindstone for sharpening hand tools.

• A pair of goggles - handy for when machine-buffing mechanical parts and when you're unsure of how electrolytic capacitors are going to react when you put the power on.

- A camera - very useful when needing to remove a number of components from an area and replace them - you can check that they're in the right place!

Tools

You can never have enough, but the one that you want is usually the one that you haven't got or can't find! For what it's worth, here is a selection from my trusty toolbag:

A diverse range of flat, cross-head, and jewellers screwdrivers; long, short, and angled-nose pliers; bradawl; small table/ bench vice; brushes, both large and small from soft to stiff, from corner to bottle type; IR thermometer; croc-clips; wire-snips; wire strippers; scalpel with round and straight blades; scissors; pens; marker pens; pencils; vernier calipers (with inside & outside capability); fibre brush; penknife; Stanley knife; centre punch; tweezers (metal and plastic); monkey wrench; a range of metric and imperial spanners and Allen keys; small brass and steel wire brushes; round and flat files; magnifying glass (hand and desk-top types); soldering irons (10 and 25W); steel rule; steel tape-measure; magnetic picker-upper; inspection mirror; small hand-torch; cyclists head-torch; feeler gauges; small electric drill/driver; paint brushes (artists; and ½" & 1" brushes to remove dust); ultrasonic bath; range of twist drills and counter-sink; solder-sucker; a few small syringes (one each for applying lubricating oil, penetrating oil, and Vaseline); and a medium hammer - or a big hammer if you're prone to getting frustrated!

Cleaning Equipment

Strong mains vacuum cleaner, plastic polish, polishing paste, isopropyl alcohol, acetone, water, air-duster, rubber gloves, dusters, chamois leather, methylated spirit, white spirit, linseed oil, glass cleaner, switch cleaner, anti-static foam cleaner, silk gloves, thick gloves (for broken glass), cotton buds, wipes. Store chemicals carefully to avoid spillages, and so that they cannot be investigated by those with little fingers!

Materials

Electrical insulating tape; wood glue; superglue; epoxy adhesive; thread-lock; freezer spray; cable ties; heat-shrink sleeving; emery paper; wire wool; washing-up liquid; scratch remover; double-sided tape; 60/40 Sn/Pb solder (or unleaded equivalent); tin of flux; solder wick. Polish for wood, plastic, and metal.

Parts

A selection or as required: capacitors, resistors, valves, ICs, transistors, diodes, switches, controls, knobs, terminals, screws, plugs, sockets, connectors, coloured wiring, screened cable, mains cable, and fuses (1, 3, 5, 13A).

Parts can be stored in labelled sets of small drawers hung from the wall. Likewise it is convenient to feed reels of wire onto a short horizontal tube or broom-handle, and to fix these on the wall so that the required colour and length can be pulled down and cut off as needed.

It is possible to waste an inordinate amount of time looking for a part – so try to label all your storage containers.

Test equipment

There is little worse than having to lug a heavy piece of test equipment, like a signal generator or oscilloscope, over to the test bench, only to find that it doesn't work properly when you need it. Therefore, your test equipment should working, calibrated, and located close to, or on, the test bench, before you need it.

As with any item of test equipment, the challenge is not just to make a measurement without damaging the instrument, but crucially, to know why the reading might not be correct. Aside from the instrument's own internal error, there are various ways that external errors can creep in depending on the circumstances, which will be highlighted in due course.

As to what you should have to hand, one

might commonly expect find an analogue multimeter; a DMM (ideally with a capacitance and inductance measurement capability); a neon mains-test screwdriver; an RF/AF signal generator; an oscilloscope; a signal injector and signal tracer; a frequency counter; a transistor and IC tester; a bench PSU; a VSWR/power meter; probes X1 and X10 for 'scopes, analysers, and generators; test meter probes, and an earth-wand (for discharging capacitors). If needed and affordable, some more expensive specialist kit such as a spectrum analyser, RF network analyser, and antenna analyser might also be on the bench. That said, you really do not need expensive kit to make the majority of measurements, and even better, there is much that can be constructed at minimal cost to add to your capability, as we shall see later.

Remote/Automated Operation

Newer instrumentation, particularly oscilloscopes and analysers, may well contain the facility to interface to a PC, smartphone, or tablet, either through a lead, or wirelessly via WiFi or Bluetooth. An application of this might be remote sensing of power and VSWR.

To use an extreme (but not entirely absurd!) example; rather than peering at the needle of a Field Strength Meter at the bottom of the garden through a pair of binoculars from the shack window, the level at the meter can now be encoded, and data sent back to the shack for easy monitoring and adjustment of the transmitter.

→ Older equipment may have a GPIB/IEE488/HP-IB 8-bit parallel bus so that a number of items of test equipment can be controlled at once. This can be a useful facility; see 'Design Notes', by Andy Talbot, G4JNT, *RadCom*, October 2016.

Environmental & Legal Considerations

There are rules covering the availability, use, and disposal of a wide variety of materials. The following should be noted:

1) Waste Electrical and Electronic Equipment (WEEE) Regulations

These are intended to control the disposal of a long and ever-growing list of materials that are considered harmful to the environment, many of which are found in electrical and electronic equipment. This means that equipment such as mobile phones, monitors, computers, and much of what appears in the average Radio Amateur's shack, should be disposed of properly, so that the waste can be recycled where possible, and taken out of use under controlled conditions if necessary.

→ For an overview see: 'The WEEE Directive and Disposal of Old Electronic Equipment', Peter Chadwick, G3RZP, *RadCom*, January 2017.

→ To locate a local facility where you can take your batteries, computers and mobile phones to, visit The European Recycling Platform website, 'Where to recycle WEEE, batteries and packaging', at: *https://erp-recycling.org/uk/where-to-recycle/*

→ Note that the UK Health & Safety Executive provide guidance on how to dispose of a list of hazard materials, including capacitors, transformers, asbestos, and mercury-containing components, at: *http://www.hse.gov.uk/waste/waste-electrical.htm#further*

→ Your local council may provide a list of items and materials, where to take them, and what you need to may do; eg call beforehand, double-bag, and label certain hazardous materials before bringing them onto the waste site.

2) Solder and the Restriction of Hazardous Substances (RoHS) Regulations

The 'long and an ever-growing list of materials that are considered harmful to the environment' mentioned above includes lead. This is bad news for the electronics industry as lead (like many of its hazardous friends), is chosen deliberately because it is the best solution in many situations. The regulations have therefore forced industry to try and find a substitute(s) that will do the job - hopefully almost as well.

Traditionally, solder for electronics has been 63/37 Sn/Pb (Tin/Lead) or around that ratio.

The hunt for a replacement has led to the development of a range of new products using various alloys in differing proportions depending on the application. The ternary eutectic compound Sn/Ag/Cu (Tin/Silver/Copper) in the rough proportions 96.5/3/0.5 seems to give acceptable performance, but unfortunately, the soldering temperature required is higher, which makes the flux vapour more toxic, and (ironically!) more hazardous for the operator. Worse still, the temperature window to make a good joint is narrower, requiring greater operator skill, plus the joint may not appear as smooth and bright compared to its leaded counterpart. There can also be issues with the long term reliability of lead-free joints - which is one reason why lead was introduced in the first place!

Having said all this, the really good news for the Radio Amateur is that he/she does not need to use lead-free solder because he/she is exempt from the regulations if working on equipment for oneself. If however, you are contemplating constructing something for someone else, whether being paid or not, you are highly likely to be required to use lead-free solder, as you are 'placing equipment onto the market'. Read the regulations and take advice before continuing!

The move towards lead-free might make leaded solder harder to obtain, so you may need to invest in a temperature-controlled iron and a fume-extractor if you don't have them already!

→ A useful primer is 'Using lead free solder', by Ian Poole,
http://www.radio-electronics.com/info/manufacture/soldering/lead_free_soldering/using-lead-free-solder.php

→ For a good overview of RoHS see:
https://www.conformance.co.uk/adirectives/doku.php?id=rohs

Reference

[1] UK Amateur Radio Licence, Section 7, 'Equipment' and Section 2(1)(1)(a-b)

2 Equipment for Measurement of AC/ DC levels and component values

2.1 The Multimeter

THIS IS a basic measuring instrument that every electronics enthusiast should have. Available as a digital or an analogue type, both sorts have their pros and cons, and so ideally you would have one of each (see **Fig 2.1**).

You can now buy a really good, brand new, multimeter of either variety for as little as £10-£20, which is fantastic value for money, although as with most things, the more you pay, the more features, better precision, and higher quality you get. For example, a basic digital multimeter will measure voltage, resistance and current. A better one could measure inductance, capacitance, temperature, and frequency as well, and might even incorporate an AF signal generator (**Fig 2.2**).

Let's look more closely at what can be measured, how, and things to look out for……

Using a Multimeter

There are a variety of parameters that can be measured depending on the model.

Voltage: A modern digital multimeter is capable of measuring a steady AC or DC voltage extremely accurately because its 'loading' effect on the circuit is normally negligible, and because it presents the result clearly on a display with large numbers. The analogue type in contrast will usually 'load' the circuit more, and should have a sensitivity of at least 20kΩ/V to reduce this effect. As the needle position can be hard to interpret, better analogue multimeters have a mirror behind the needle so that they can be read 'head on' (ie with no parallax) to minimise the scale interpretation error (**Fig 2.3**). That said, the analogue multimeter is often superior where the voltage is moving, as fluctuations can be followed far more easily. For the repair of valve equipment, I would prefer the latter for voltage measurements for the reasons

Fig 2.1 An analogue and a digital multimeter side-by-side

given, and additionally because if there is AC (mains hum) on the DC (HT) supply, a digital multimeter can get very confused and give

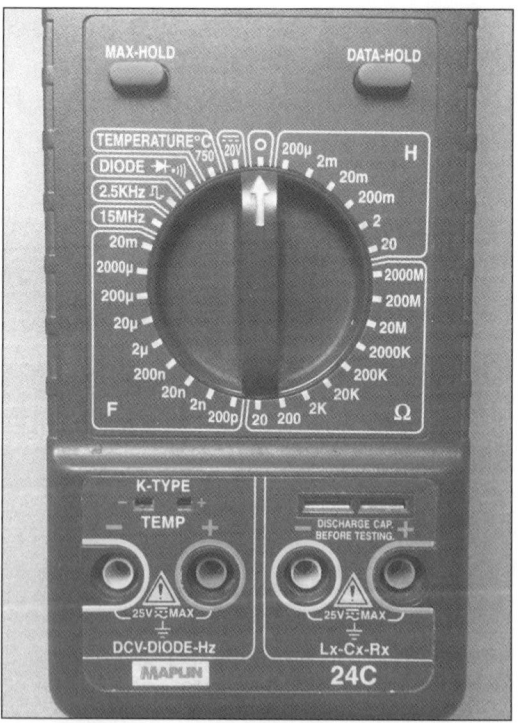

Fig 2.2 Close-up of functions available on a typical DMM

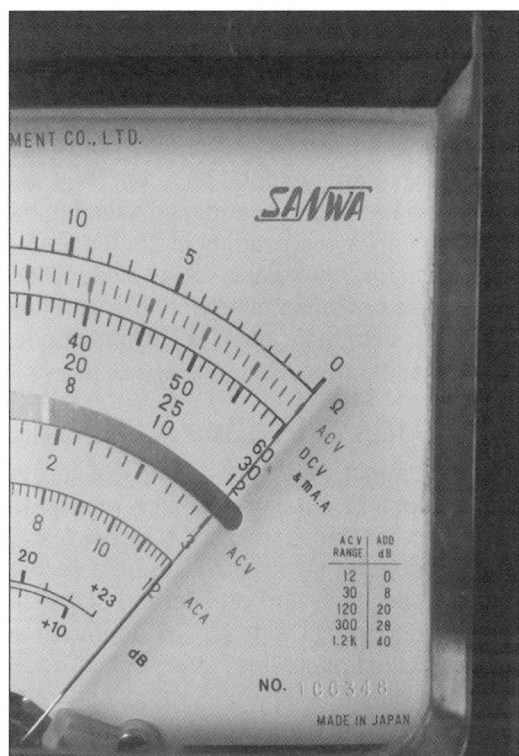

Fig 2.3 Avoiding parallax error with a mirror behind the scale an analogue multimeter

into its own here and can give a quick and accurate result. However, when measuring large resistances (say >1MΩ) ensure that the multimeter's internal resistance (shown in its specification) is at least 10x the resistance that you're trying to measure (and preferably a lot more) to minimize shunt errors. Cheaper multimeters may have an internal resistance of only 10MΩ, the best ones in the order of 10GΩ.

CARE! Never attempt to measure resistance with the power on, and ensure that one end of the resistor is disconnected from the circuit to avoid a false reading caused by other components shunted across it. To avoid errors when measuring low resistances, ensure that your probes are clean and well-fitted at the other end. Bringing the two probes together, an analogue multimeter should go roughly full-scale: use the 'zero' knob to manually set the pointer to precisely '0 ohms' on the dial. Make good use of the parallax mirror if provided. A digital multimeter should auto-zero, or show a small resistance (<1Ω) that ought to be deducted from any low resistance measurement.

Inductance and Capacitance: Connect the isolated component to the probes or slots provided. Remember that any test leads inherently have a small self-inductance along them and a small self-capacitance across them. This 'parasitic' value should be noted and subtracted from the item being tested if it significant in comparison.

Temperature: Sometimes a thermocouple on a twisted-pair cable is provided which plugs into the multimeter. Note the thermocouple temperature range, and secure with tape, and possibly heat-sink compound, to the item to be measured. If you use a different thermocouple, make sure that it has the right type letter (eg 'J'), otherwise readings will be incorrect.

Frequency measurement: An in-built frequency counter may allow limited in-circuit frequency measurement via the test probes. Do not exceed the maximum specified voltage between the probes when connecting to

a variable and misleading readout. Beware also, that the working range of many digital multimeters may extend to only to a few tens, rather than hundreds of volts, because they are intended to test modern transistorised equipment, and so are not suitable for the task. Either way, when you pick up those two probes, know the highest voltage likely to be encountered and set the multimeter to a suitable range *before* making a measurement. If you don't, you may trigger the 'quick-blow' fuse, damage the instrument, and (if analogue) permanently bend the needle or burnout the movement. Also be aware that the digital multimeter contains a clock oscillator to process the display, and the pulses may appear as interference transmitted through the test probes.

Current: Similar arguments apply - ideally the working range should extend to at least a few amps.

Resistance: A digital multimeter really comes

a powered set in this mode. Normally intended for probing transistorised equipment, you may need an external capacitor (DC block) with a high DC working voltage in order to protect the multimeter.

Signal-Generator: This may usefully provide a basic AF waveform (perhaps sine, square, triangular, or sawtooth) to the probes. Again, do not exceed the maximum specified voltage between the probes when injecting a signal into a powered-up set; use an external DC block if necessary.

Handling a nice multimeter is one of the joys of electrical construction and maintenance. If you understand how it works and use it correctly, your joy will be complete, and you will have mastered a basic competence essential to any electrical work.

→ To understand more about how the multimeter works, study Appendix A.

→ For occasions when there is a need to monitor the applied voltage, current drawn, and perhaps ripple on the supply to a test piece, it may be convenient to have a simple 'breakout box' that can be inserted, rather than tie-up more than one multimeter and have a nest of wiring. Other functions can be added. The following shows how to make one: 'Multimeter Breakout Box', Geoff Theasby, G8BMI, *RadCom*, May 2017.

→ A cheap and simple way to check the calibration of a DVM is to use it to measure a precision voltage reference. Such a reference may be made or purchased for just a few pounds. For an example using an AD584 IC, see 'Homebrew', by Eamon Skelton, EI9GQ, *RadCom*, October, 2016.

2.2 The Multitester

One can now buy a microprocessor-based hand-held tester for £10-20 (or a kit to make one for half-that) which can accurately measure the properties of many active and passive components. Not just inductance, capacitance, and resistance (and typically over a wider range than a multimeter), but capacitor ESR and diode and transistor characteristics

(polarity, leakage, barrier potential, gain etc) as well. It may also check the operation of some common ICs - make sure that it covers the ones you're interested in. They are so clever that they will even work out which leg is which, so you don't have to! They are quick and easy to use, and give valuable information on a wide variety of component – very handy if you have a lot of unidentified components of uncertain history to test! They represent extremely good value for money, and are another 'must' for the shack (**Fig 2.4**).

2.3 Leads & Probes

These connect the circuit under test to the test equipment. They need to be robust (to withstand constant use), should be quick and easy to attach, have minimal impact on the circuit under test, and present the signal to the test equipment with as little degradation as possible. Ensure they are kept in good condition and are adequately rated, especially if working with high voltages or currents.

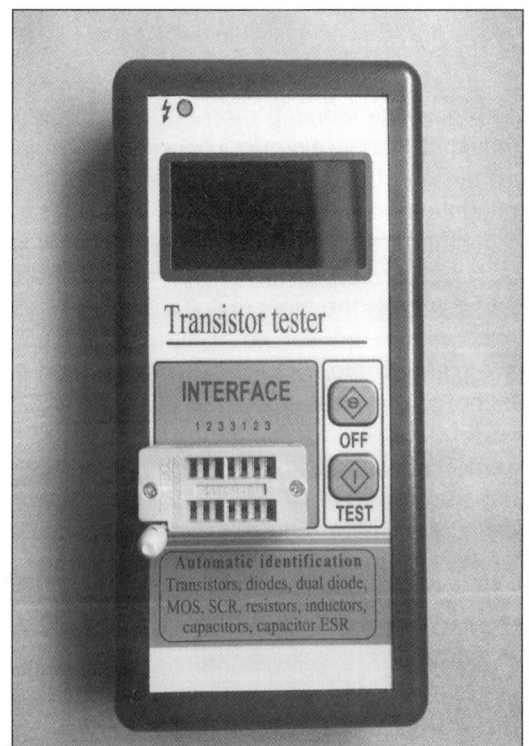

Fig 2.4 A typical multitester

DC/LF leads

Typically attached to multimeters and the like, these are characterised by having a push-fit plug at the test instrument end, and either a probe point, or a press & release grip to the equipment under test at the other. These are connected by a length of flexible, multi-stranded, insulated copper wire. With use, the torsion can gradually break the strands, causing an open circuit. Depending on the design, it is often possible to remove the plug, probe, or grip cover to re-solder the joint, which should be done carefully and neatly so that joint resistance is minimised, the bond is strong, and the solder does not foul the cover when you try to reattach it (**Fig 2.5**).

Always try to keep the probes clean, as any dirt or grease will act to increase resistance

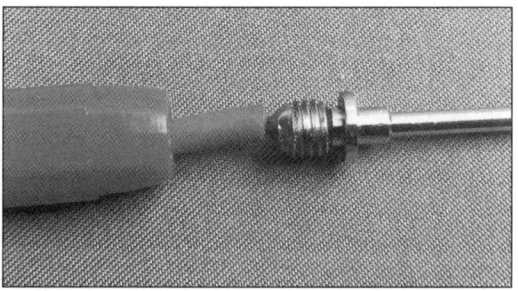

Fig 2.5 When re-soldering the probe to its cable – make a neat, strong, and compact joint like this

causing measurements to be inaccurate and erratic. Some typical leads and probes are shown in **Fig 2.6**.

RF Leads & Probes

At low frequencies, a coaxial lead terminated in two crocodile clips may suffice for the injection or monitoring of a signal (**Fig 2.7**). One of the crocodile clips is connected to the centre conductor of the coax, whilst the other is on a short length of flexible wire connected to the braid. The braid connection is the earth, whilst the actual signal being examined is connected via the centre conductor. This forms a very basic (one might say crude) probe, which suffers from the fact that the crocodile clips are large and clumsy and often uninsulated. Beware of damage to the circuit by the clips shorting-out, and to you, if there is a large voltage present!

If connected to an oscilloscope with a 1MΩ input impedance, the input impedance presented by the circuit is typically in the order of 1MΩ, shunted by about 60pF or more due to the cable and the capacitive nature of the flying leads. Handy for quickly and simply injecting or monitoring a signal in low frequency work, this type of probe is unsuitable for accurate RF measurements because of its poor frequency response.

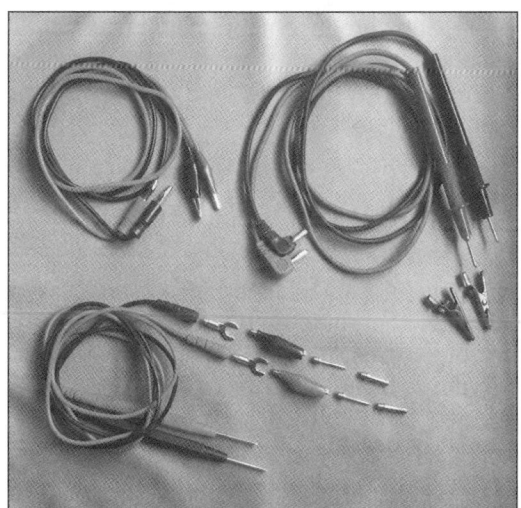

Fig 2.6 A selection of typical DC/LF test leads and probes

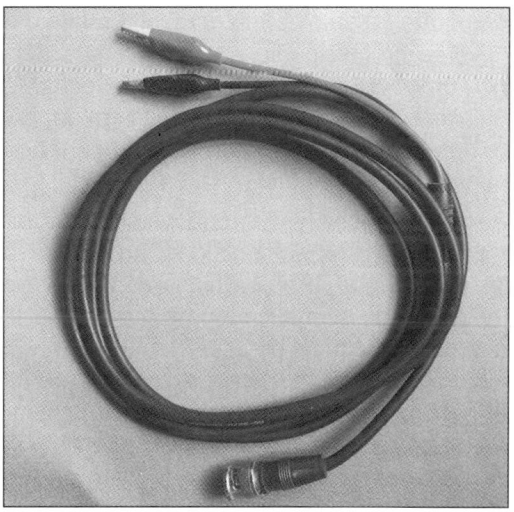

Fig 2.7 A low-frequency RF test lead / basic probe

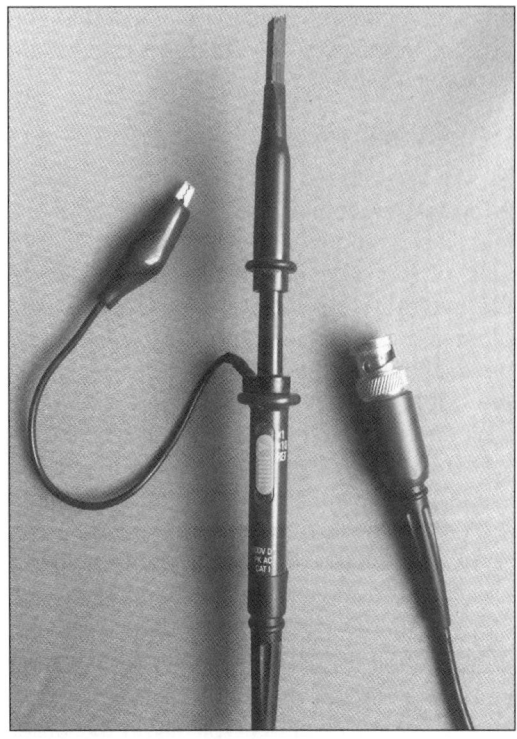

Fig 2.8 A 'proper' RF probe

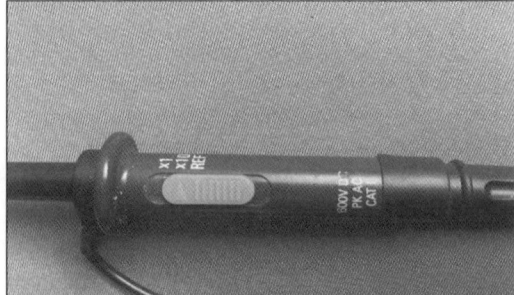

Fig 2.9 Close-up of the RF probe's switch positions and its low frequency voltage rating

A much better arrangement is a lead with a calibrated probe (**Fig 2.8**).

The probe is usually plastic covered (ie insulated), has several attachment options available to connect to the circuit, and incorporates a small earth lead with miniature covered crocodile clip. A typical 'times one' (x1) probe as it is called, when connected to an oscilloscope with a 1MΩ input impedance, will present the circuit with an impedance of about 1MΩ shunted by about 40pF.

A variant on this type is one which has a divide-by-ten arrangement in it, known as a 'times ten' (x10) probe, which attenuates the input voltage by a factor of 10 and so allows a larger value to be measured. Importantly, it also raises the impedance across the circuit by ten times (to 10MΩ in this example) shunted by less capacitance (about 15pF), both of which helpfully decrease the loading on the circuit being measured. A more expensive probe will have a slide-switch which allows

selection of either x1 or x10 and may also provide a 'reference' position which grounds the probe, and is useful for calibration purposes when making a measurement (**Fig 2.9**).

The probe will often be useful to 60 or 100MHz and should come with other information, such as its input capacitance and voltage derating with frequency.

An RF probe will often incorporate a 'compensation' trimmer which can be adjusted for best frequency response (**Fig 2.10**). A plastic adjuster may be provided so that the capacitance setting is not influenced by a metal screwdriver.

→ For more information on compensated probes and oscilloscopes, see Chapter 3.

Again, probe leads need to be keep be kept clean, and care taken not to strain the cable, especially at the two ends where it enters the housing, as damage and poor measurements will result.

As to interconnecting equipment, for HF work, these leads are normally coaxial (for low-loss RF transmission) and will usually have BNC plug at each end to connect, for example, a signal generator or an oscilloscope to the equipment under test. At higher frequencies, this may change to perhaps N-type, or SMA.

Logic Probes

These help to investigate the operation of digital ICs and associated circuitry. They will normally detect logic hi-, lo-, and indeterminate state, and display the result by means of a green, red, and amber LED respectively, and/or with audible support. They may have a switch to swap from continuous running to 'single-shot' mode, and another to change from TTL to CMOS. Contact with the circuit under test is via a pointy metal probe, and power may be drawn through a couple of wires which come out of the body of the probe and terminate in crocodile clips so that they can be connected to the digital equipment's ground and supply rail. They may also have facility for internal power. Such probes are inexpensive to purchase (costing around £10 upwards), and simple to make, perhaps being housed in the shell of a pen.

→ Search the multitude of circuit and construction ideas on the internet for how to make one.

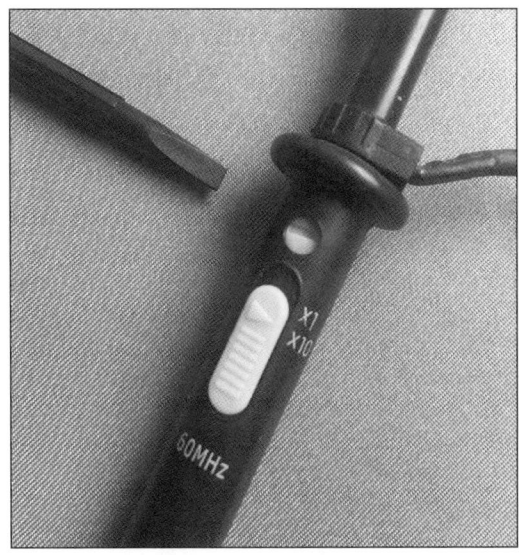

Fig 2.10 Close-up of the compensation screw and its non-metallic adjuster

RF Diode Probes

Measurement of RF voltage can be problematic as multimeters are not designed to measure at radio frequencies, and many oscilloscopes struggle beyond perhaps 20 or 30MHz. An alternative is to make a simple RF detector, comprising a diode, capacitor and resistor housed in a slim metal cylinder with a metal probe to the circuit and a coaxial output to a moving-coil meter or a digital voltmeter. If more sensitivity is required, the RF can be amplified before detection by an IC.

→ To build your own, see Chapter 8.1 'Probes'

2.4 Other Kit

The ability to make quick and accurate measurements of inductance and capacitance with a relatively inexpensive hand-held multimeter or multitester has probably been responsible for the demise in the use of test gear based on resonance of a tuned-circuit to determine these and related parameters. For example the:

• LCR meter (for Inductance, capacitance and resistance measurement)

• Q-Meter (for measuring the quality factor of a tuned-circuit, and for making capacitance and inductance measurements.

• GDO (Grid-Dip Oscillator - or its solid-state equivalent using an FET – for determining the frequency of a tuned-circuit, and which may double as an absorption wavemeter)

However, they can be a useful asset if obtained cheaply and in good condition.

→ To build a thoroughly modern version of the GDO using a microprocessor and LCD display, see: 'Build a Digital Dip Meter', Stuart Ball, Circuit Cellar, February 2013, Issue 271. *http://circuitcellar.com/wp-content/uploads/2013/01/CC25_ProjCard_Pres_Ball-CC271.pdf*

3 Equipment for Time & Frequency Domain Analysis

3.1 The Oscilloscope

THE OSCILLOSCOPE (**Pic 3.1**) is an extremely useful piece of diagnostic equipment to have in the workshop, not least because it has a pictorial rather than a numerical display, which allows signs of distortion, noise, DC offsets, and modulation on a signal to be easily identified. The instrument permits the amplitude of a signal, and it variation, to be easily measured, and despite working in the time domain, it is usually possible to at least estimate the frequencies involved.

A basic cathode ray oscilloscope (CRO) with a bandwidth to a few tens of MHz can be bought second-hand for just a few tens of pounds, and can be ideal for AF, IF, or low-frequency RF work. Alternatively, you could purchase a modern LCD version for a few hundred pounds, or a small box which will convert your PC into an oscilloscope for a price somewhere in between the two. Digital oscilloscope kits are available for just £10-30 online, although you should check that the quality and functionality is adequate for your needs before you buy.

The modern equivalent to the old CRO is more expensive, and is likely to have all sorts of 'bells & whistles' – the question is: do you really need them? There is a lot of satisfaction to be had from skilfully using a cheap CRO; however, unless you are prepared to maintain it (for example deal with an electrically noisy y-shift control) a modern version might better.

Operation

Fig 3.1 shows a block diagram for an oscilloscope reduced to its four fundamental elements:

1) A power supply: to suit the needs of its amplifiers, timebase, display, and other internal electronics.

2) A Y-amplifier: The signal of interest (whose amplitude could range from mV to tens of volts) is applied to the Y-channel of the os-

Pic 3.1 A cheap second-hand dual-trace CRT oscilloscope – old, but perfectly good for many test measurements

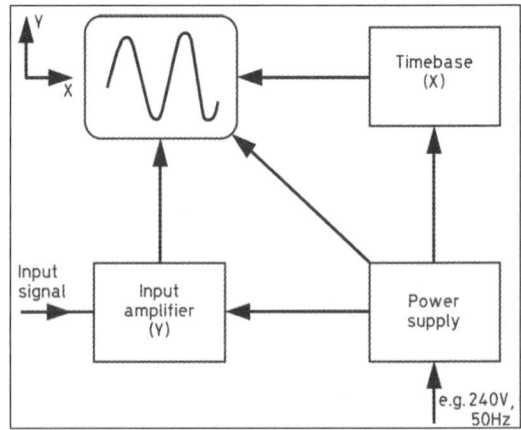

Fig 3.1 Block diagram of an oscilloscope

cilloscope and amplified to produce a vertical movement on the screen. This Y-deflection therefore represents the signal's amplitude at a point in time, so to facilitate measurement on the vertical axis, the amplifier's Y-gain is made variable by a control calibrated in V/ division or mV/division.

3) A Timebase: An oscillator which usually drives an X-amplifier to cause the waveform to traverse the display in the horizontal plane. At the end of the transit, the spot is made to return (flyback) to the starting position ready for the next display without being seen (blanking). The horizontal X-axis deflection represents time, and so the speed of the transit is made variable by a control calibrated in s/division, ms/division and μs/division.

4) A Display: There are three types, which, at the end of the day, tend to determine the shape and size of the overall equipment:

Cathode Ray Tube (CRT)

With a single-trace CRO, an electron-gun fires a beam of electrons at the phosphor coating on the inside of the glass screen, causing it to illuminate. The focus, brightness, and astigmatism controls set electro-potentials in the CRT and thus enable the user to form a small, bright, round spot that can be deflected vertically by the signal applied to the Y-amplifier, and horizontally by the signal from the timebase and X-amplifier.

In a dual-trace CRO, control of this beam is split in turn between the two Y-channels, giving two traces from one beam.

Less common is the dual-beam CRO which utilises two independent electron guns and therefore permits comparison of traces driven from two separate timebases.

CROs are generally analogue, with a few specialist units (such as sampling 'scopes) using digital techniques. Unless the display is continually refreshed, or there is persistence (storage) the trace will just fade away.

Horizontal and vertical measuring divisions are usually marked on a clear plastic sheet (a graticule) placed in front of the display.

Liquid Crystal Display (LCD)

These types generally use an ADC (analogue-to-digital converter) to convert incoming signals into a digital value which can then be stored in RAM and used to plot the signal on a display under microprocessor control. Each trace may be a different colour so that the signal in each channel can be easily identified. Most of these can easily store waveforms as well as communicate directly with a computer for further storage, printing, and analysis. They may also have digital

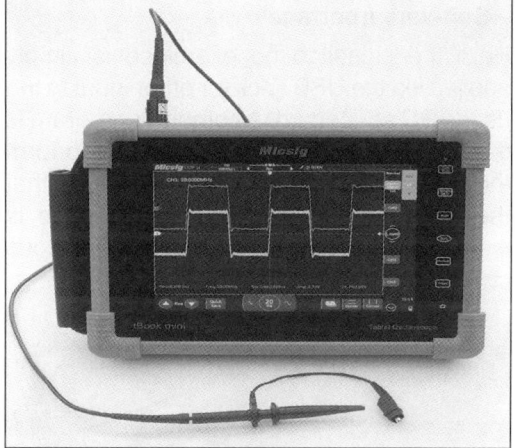

Pic 3.2 A modern touch-screen oscilloscope

voltmeters and frequency counters incorporated into them that will show digital values on the display as well.

A modern tablet-like touch-screen oscilloscope **Pic 3.2** may boast a bandwidth of 100MHz, and the ability to display at least four bright and colour-coded channels. Moreover, it may automatically calculate and display the parameters of common waveforms and have the capacity to transmit results wirelessly, or via Ethernet LAN, to a remote location.

→ For a detailed review of a modern commercial oscilloscope product see:

'Microsig tBook mini Oscilloscope', Mike Richards, G4WNC, *RadCom*, March 2018.

Personal Computer (PC) Screen – Hardware Interface

These types also use a high-performance ADC to convert the incoming signals into a digital value, but rely on the power and functionality of the ubiquitous personal computer to do the processing and display the output. They consist of a box which interfaces to the PC through a USB or similar transport cable, and usually connects with the circuit under test using compensated probes (**Pic 3.3**).

Personal Computer (PC) Screen – Software Interface

Here the signal to be examined is simply applied via the USB, jack, or other input to the user's PC soundcard. A computer program is then run which processes the signal to form the oscilloscope display. Note however, that the bandwidth of this 'soft' oscilloscope is limited to that of the PC sound card. Various

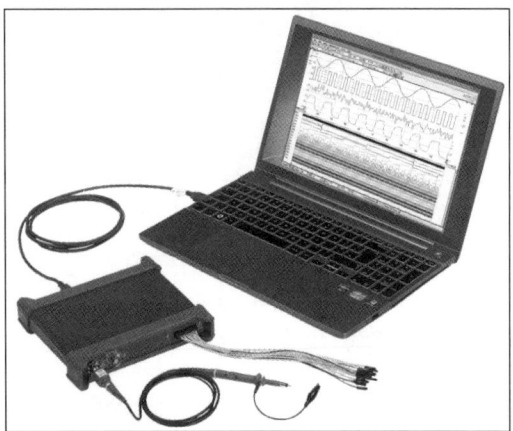

Pic 3.3 A modern PC oscilloscope

programs are available which can perform this function, and they may provide other facilities such as audio-spectrum analysis. For a long list of options covering Windows, Linux, Mac, OS, and Android platforms see: *www.dxzone.com/catalog/Software/Oscilloscope*

Two examples from the above are:

BIP oscilloscope: Windows freeware; it automatically uses the highest sample frequency

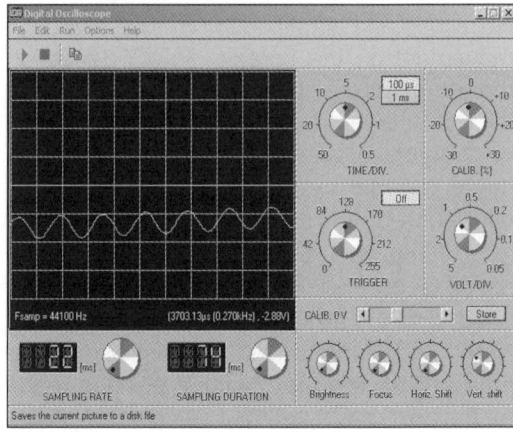

Pic 3.4 BIP oscilloscope display

available at the input that is selected (usually maximal 44kHz). See **Pic 3.4** for a screen shot.

Accuracy: The oscilloscope uses 8-bit samples to read the input signal.

Input impedance: The same as the input impedance of the sound card.

Xoscope: Linux open source freeware. This is an 8-channel oscilloscope for audio display, menu-driven rather than by graphical buttons. A manual is available from the Help menu. It can function with add-ons such as Bitscope and Probescope. See **Pic 3.5** for a screen shot.

Care! You may need to isolate your PC soundcard from harmful DC levels generated by the equipment under test (and vice-versa) by using a 1:1 audio transformer or other arrangement. Also, if using a PC output as a control signal you may need to insert an optical isolator for similar reasons.

Controls

The following are typical:

• Y-Gain (volts/div): changes the gain of the vertical amplifier so that a suitable size signal can be displayed on the screen.

• Y-position: allows position of the trace to be varied vertically in order to place it at a convenient point on the screen.

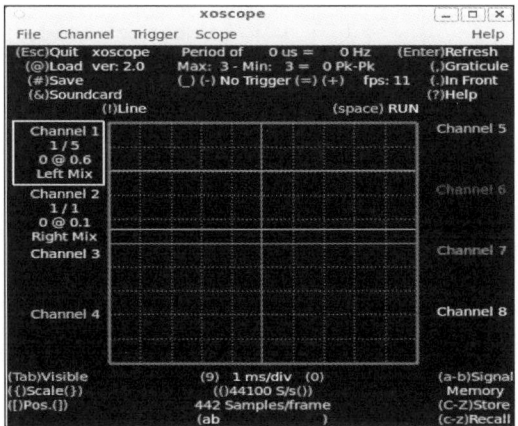

Pic 3.5 Xoscope display

- AC/DC/GND: selects the type of coupling to the input signal. 'DC' allows the AC signal plus any DC to be displayed; 'AC' removes the DC component from display; 'GND' allows input to be grounded internally in the oscilloscope in order to zero-position the trace.
- CH1/CH2/ALT/CHOP/ADD: selects how the channels will be displayed. Channel 1 or 2 separately; both channels on an alternate basis; both channels, but chopping between them during a single scan; Channel 1 and 2 added together - there may also be a channel-invert function, which allows subtraction instead.
- Timebase (Time/Div): changes the speed at which the trace traverses the screen and thus affects the number of cycles of an input signal being displayed on the screen.
- X-position: allows horizontal shift of the trace in order to put it in a convenient position for taking measurements.
- X-magnification: normally allows a x10 magnification if required, equivalent to a faster time on the timebase.
- Trigger Level: determines the trigger level of the input signal. This can be put to AUTO or set to a given level. The input signal triggers the timebase so that a stationary waveform appears on the screen.
- Source: selects the source for the trigger circuit: Internal, Channel 1, Channel 2, Line (mains) or External.
- Intensity: controls the brightness of the image displayed on the screen.
- Cal: an output for X- and Y- amplifier calibration purposes, usually at a frequency of 1kHz with a fixed amplitude, often 0.1V peak.
- CRT displays will usually have a Focus, and possibly Astigmatism and Rotation controls as well to adjust the sharpness, shape, and orientation of the image respectively.
- LCD and PC oscilloscopes usually have the added capability of automatic measurement functions, which can make the determination of key parameters such period, frequency, peak-to-peak voltage, and waveform rise-and-fall times so much easier. These measurements are displayed on the screen in real-time, perhaps with the option to average or hold the value. As well as enabling quick display of the key characteristics of different types of waveform, there are usually markers that can be placed on the waveform, so that parts of it can be investigated and measurements taken.

Compensated Probes

Although one may use a common (uncompensated) probe (Chap 2.3) to connect the oscilloscope to the circuit under test, its bandwidth is significantly limited by its internal resistance. A dedicated oscilloscope probe in contrast, incorporates a parallel capacitance which can be adjusted by a screwdriver to achieve a flat response in the bandwidth of the oscilloscope (**Pic 3.6**).

This is normally achieved by connecting the compensated probe to the oscilloscope's 'Cal' output and adjusting the probe's compensating capacitor until an undistorted square-wave is obtained (**Pic 3.7**). **Fig 3.2** shows possible responses when the probe is connected to the oscilloscope's calibrator jack.

The coaxial lead on a compensated probe should not be cut, as it may not be possible to compensate for the new length, and reflections caused by impedance mismatch may also arise.

The shortest ground lead possible should be used from the probe to the circuit ground, as long ground leads are inductors at high frequencies, causing 'ringing' and other undesirable effects.

Voltage Measurements

The easiest voltage measurement to take is the peak-to-peak value. For example in **Pic 3.8**. If the vertical displacement (Y) is six divisions from peak-to-peak and the Y-gain is 0.5V/div, then the peak-to-peak voltage is 6x0.5=3V. The peak value is half of this (=1.5V), and the RMS value is $1/\sqrt{2}$ (or 0.7071) times this value (=1.06V).

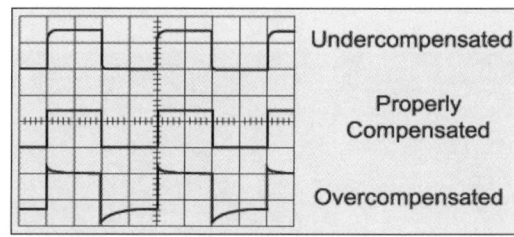

Fig 3.2 Under-, over- and properly compensated probes (reproduced with kind permission of the ARRL)

Accuracy of voltage measurement is unlikely to be better than about 5-10% when taking readings from a CRO trace. However with LCD and PC equipment, the on-screen displayed value will be much more precise.

Time and Frequency Measurements

For a sinewave, it is best to measure from like-point to like-point, such as between two adjacent negative peaks. For example, in **Fig 3.3** this distance is eight units. If the X or time-base setting is 0.5ms/div, this represents a period of 8x0.5=4ms. The frequency is the reciprocal of this, ie 1/4ms or 250Hz.

The same principle applies to a rectangular waveform, whose edges are clearer, and whose amplitude, period, and frequency are much easier to estimate. For a pulse waveform, another parameter that can be measured is the mark-to-space ratio, ie the ratio of time when it is high to when it is low. In the case shown (**Fig 3.4**), it is a measure of the durations on the screen and is 4:2 (which

Pic 3.6 Adjusting a compensated probe

Pic 3.7 A correctly compensated probe

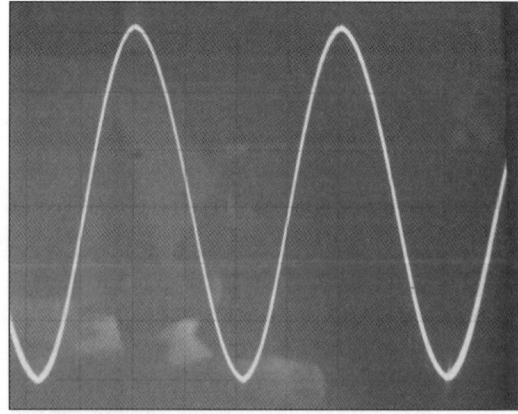

Pic 3.8 Measuring a sine wave

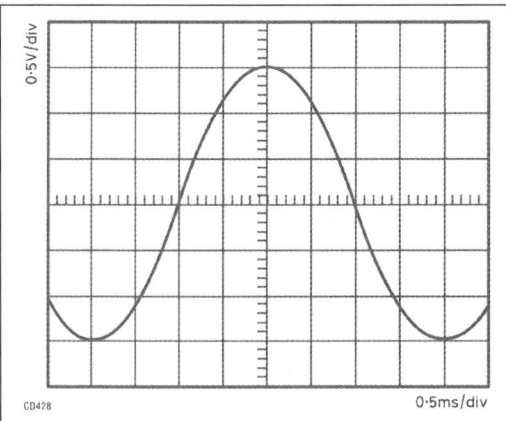

Fig 3.3 Measurement of time and frequency

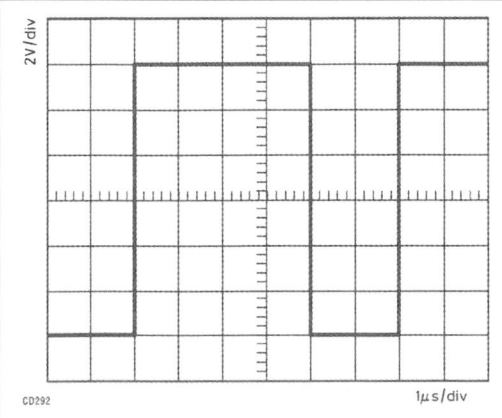

Fig 3.4 Measurement of mark-space ratio and duty cycle

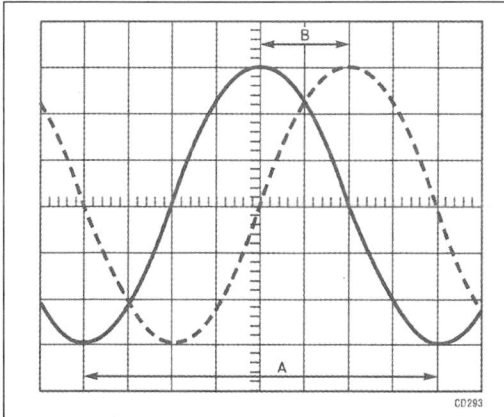

Fig 3.5 Phase measurement – first method

is the same as 2:1). The duty cycle can also be ascertained, which is the ratio of it being high to the duration of the whole pulse cycle, in this instance 4:6, ie 2:3 or, expressed as a percentage, 66.7%.

Differential Phase Measurement

There are a couple of methods that can be used here. The first (and probably most common) allows two signals of interest (which we shall call 'A' and 'B') to be displayed simultaneously together, with the associated phase difference. This method can cope with any shape of waveform, the upper frequency limit being dependent on the oscilloscope and any small phase shifts introduced by the two input-channel amplifiers (Y1 and Y2).

Signals A and B are applied to Y1 and Y2 respectively and the gains adjusted so that the waveforms are approximately equal - **Fig 3.5** shows a typical display. For most accurate results, the input selection is set to CHOP. The phase shift can then be calculated from the amount of offset between the traces. Remembering that one cycle is equivalent to 360°, the phase shift (in degrees) = 360*B/A.

The second method is most suitable for the audio frequency range or just above. It assumes a sine wave input, and only requires a single-channel (rather than dual-channel) oscilloscope. This method relies on one signal being applied to the X-input - see **Fig 3.6**.

To start with, both channels (X and Y) are simultaneously connected to the input signal and adjusted so that a suitable sized display is obtained and there is a straight line at approximately 45°. Next, the Y-channel is connected instead to the output signal of the circuit under test. Various displays may be obtained.

Fig 3.7 shows how these can be interpreted and the phase-angle obtained.

Differential Voltage Measurements

Occasionally one may have to make a voltage measurement between two points in a circuit, neither of which is grounded; for example the voltage V_{AB} in **Fig 3.8**. This can

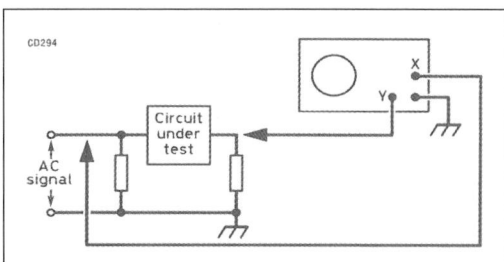

Fig 3.6 Phase measurement - second method

be accomplished in two ways. The first is to measure A with respect to ground and then B with respect to ground. The difference between these two values is equal to V_{AB}. This is satisfactory for DC and if any AC waveforms at A and B are identical in shape. It is also the only method available for a single trace oscilloscope.

The second method is applicable for both dual-trace and dual-beam oscilloscopes. There will be controls on the oscilloscope that will allow display of Y1+Y2 and Y1-Y2. In the latter case, one must use the invert control on channel Y2 and then the ADD facility; this will then give Y1- Y2. Make sure that the gain settings on both Y1 and Y2 are the same, the Y1-Y2 display is set up, and for the dual-beam oscilloscope only: that both timebases are set identically. Put one probe (Y1) on A and the other (Y2) on B; the display will then show V_{AB}. The magnitude and time readings are then as described earlier.

Equipment Limitations

Things to watch out for:

Frequency response: This is limited by the Y-amplifier (plus CRT) bandwidth and by sampling rates. Particular care must be taken with rectangular waveforms, as these consist of a series of harmonics in fixed ratios, and the higher harmonics will exceed the oscilloscope's bandwidth. The effect of this is that the displayed waveforms will become progressively more rounded as their frequency

rises, eventually tending to a sinewave.

Loading: Where the capacitance presented by the oscilloscope probe (perhaps 20-40pF) is of the same order as the circuit under test, the circuit will be disturbed and may be especially noticeable with tuned circuits, as their frequency $f = 1/(2\pi\sqrt{(LC)})$. A divide-by-10 probe will help to reduce loading effects on the circuit, but reduce (rather than eliminate) is the operative word.

Safe Input Voltage: The oscilloscope's maximum input voltage is typically 400V DC plus the peak AC signal that can be displayed. Exceeding this will damage internal components of the oscilloscope. Although a divide-by-10 probe can be used to extend the voltage range, these have a voltage limit

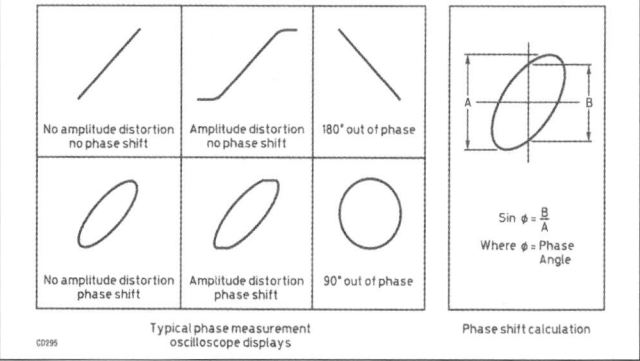

Fig 3.7 Displays and calculation of phase shift

(typically 600V DC), but this may need to be de-rated as frequency rises - check the specification. For most semiconductor applications the voltage limit does not cause a problem, however for high-voltage valve circuits one must pay due regard to these limitations for one's safety.

→ For a good

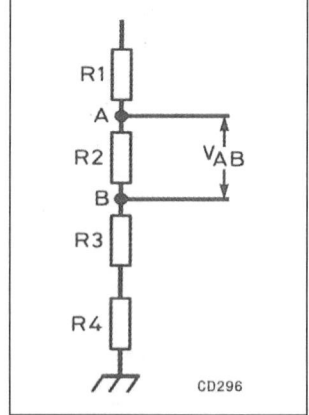

Fig 3.8 Differential voltage measurements

tutorial on oscilloscopes (including compensated-probe adjustment), and with links to further information, see:

https://learn.sparkfun.com/tutorials/how-to-use-an-oscilloscope

http://www.radio-electronics.com/info/t_and_m/oscilloscope/tutorial-basics-introduction.php

3.2 The Spectrum Analyser

Unlike the oscilloscope, which displays the signal in the time domain, the spectrum analyser presents it in the frequency domain, which is extremely helpful for assessment of factors such as modulation, intermodulation, noise, and harmonics. By addition of a tracking generator, the spectrum analyser can also display the response of filters and networks.

Commercial Products

Historically a very expensive item, they are rapidly becoming more affordable as second-hand CRT-based models increasingly come on to the market, and as LCD software-based models (either stand-alone or, using an external PC or oscilloscope for processing and display) take over. **Pic 3.9** shows a typical spectrum analyser.

→ To build your own see:

'SA3000 100kHz to 3GHz Spectrum Analys-

er', Rubens Fernandes, VK5FE, *RadCom*, August 2018.

'The simple man's Spectrum Analyser'

http://www.hanssummers.com/spectrumanalyser.html

Operation

There are various types of spectrum analyser.

Self-Contained Swept-tuned

This is the most common and sweeps a desired frequency range sequentially in time to display its results on a screen. Good for displaying periodic and random signals, it fails to capture transient responses.

Self-Contained Real-time

Real-time analysers sample the whole frequency range simultaneously and thus preserve the time dependency between signals as well as the capabilities of a swept-tuned type.

Hardware Interface - PC/Oscilloscope Screen

A potentially cheaper and more flexible alternative to a self-contained spectrum analyser is to use an add-on unit (a hardware interface) which uses an oscilloscope or PC as the display medium. Such units have a lower frequency limit (eg 400kHz), and an upper limit being dependent on type, eg 100, 250, 500MHz. They come in a range of profiles, from a large probe, to a stand-alone box. The bandwidth of the oscilloscope or PC is not usually a limitation as the high-frequency processing is taken care of within the add-on unit.

Software Interface - Personal Computer (PC) Screen

As with the 'soft' oscilloscope, the signal to be examined is simply applied via the USB, jack, or other input to the user's PC sound-card. A computer program is then run which processes the signal to form the spectrum analyser display. Again, the frequency limit of the spectrum analyser display is that of the PC sound card, creating an analyser for the audio spectrum only. However, that limitation could be overcome by adding a down-converter to baseband, which would enable a

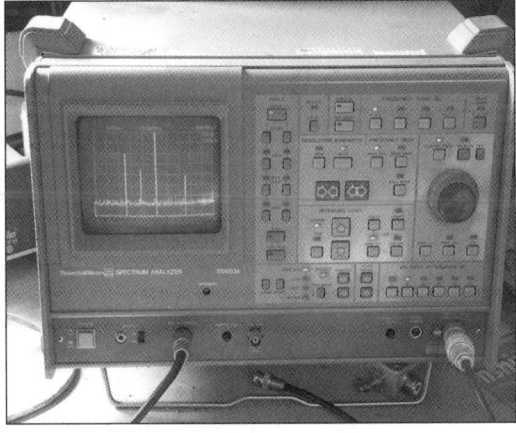

Pic 3.9 A typical spectrum analyser

20kHz or so slice of spectrum about a carrier or harmonic to be examined, for example when performing a two-tone test. A block diagram for such a system is shown in **Fig 3.9** .

→ For software-based spectrum analysers covering most operating platforms, see a long list at: *www.dxzone.com/catalog/Software/Spectrum_analyzers*

→ An example from the above site is 'Spectrum Lab', Windows, freeware by DL4YHF. A typical display is given in **Pic 3.10** It features include the ability to:

• Analyse the spectrum of an audio signal via the PC's sound card;

• Analyse the spectrum of a previously recorded sound in a wave-file;

• Observe how the spectrum changes over the time, by means of a 'waterfall' display;

• Perform n-th order audio filtering in real-time, output sent to the sound card again;

• Generate and decode some 'special' digital amateur radio communications modes such as MTHELL QRSS, DFCW, PSK, MSK.

General Use

The horizontal axis is linearly calibrated in frequency with the lower frequency to the left (which is never zero because of the way the analyser works!) and the higher frequency to the right-hand side of the display. The operator can either choose these upper and lower frequency limits, or alternatively scan over a range centred about a specific frequency.

The vertical axis is amplitude and can be either linear (V/div) or logarithmic (dB/div). For most applications, a logarithmic scale is chosen (typically 10dB/div) because it enables signals over a much wider range to be seen on the spectrum analyser. As the scale is normally calibrated in dBm (decibels relative to 1 milliwatt) absolute power levels as well as differences in levels can be seen and measured. **Fig 3.10** shows a typical display at 10dB/div vertically and 1MHz/div horizontally.

A spectrum analyser will typically be rated for a maximum input power of +20dBm (0.1W),

beyond which damage to the instrument will occur. In a similar manner, there will be a maximum permissible DC input voltage.

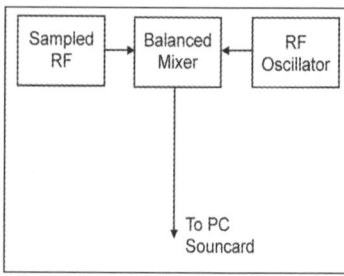

Fig 3.9 Typical arrangement for PC soundcard spectrum analyser

→ For a good tutorial on spectrum analysers see:

http://www.radio-electronics.com/info/t_and_m/spectrum_analyser/rf-analyzer-basics-tutorial.php

→ See also a 'Guide to Spectrum and Signal Analysis': *https://www.anritsu.com/*

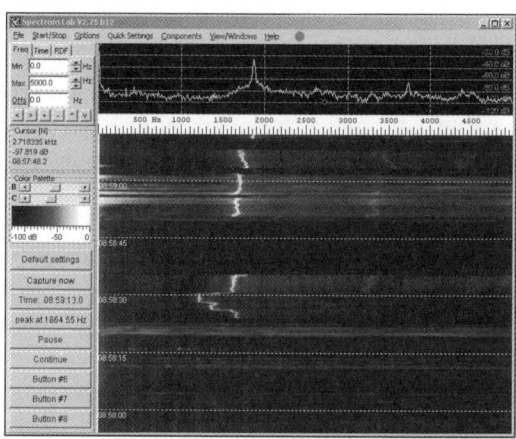

Pic 3.10 Typical Spectrum Lab display

Measurement of Harmonics

Fig 3.11 shows a typical display and depicts a 30MHz signal with harmonics at 60, 90, 120 and 150MHz. The third harmonic (for example) is shown 48dB down on the fundamental but the display does show outputs at the 2nd, 4th and 5th harmonics as well. If the coupler and spectrum analyser had been adjusted so that 0dB represents 100W, then the third harmonic power content is -48dB down relative to 100W or 1.6mW. Measurement of the harmonic output of a transmitter is the obvious application here.

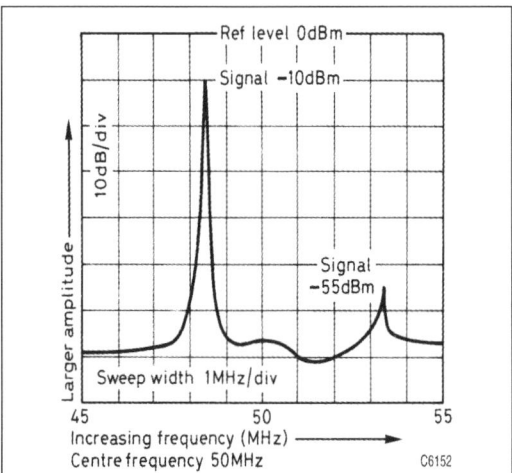

Fig 3.10 Typical screen display of a spectrum analyser

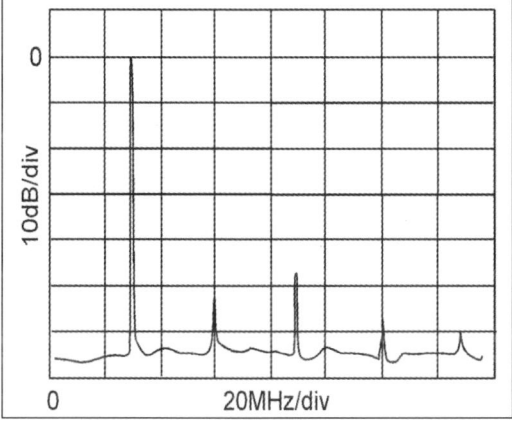

Fig 3.11 Spectrum analyser display showing harmonics

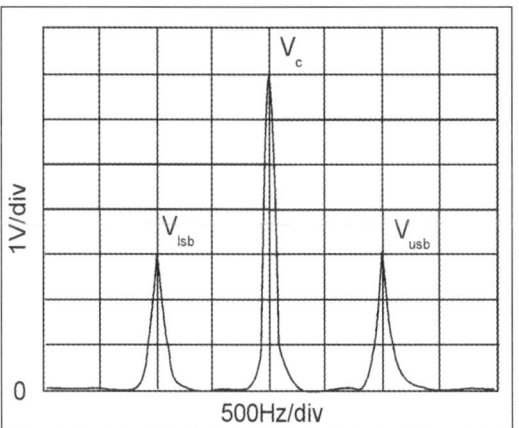

Fig 3.12 Modulation depth measurement using a spectrum analyser

Measurement of Amplitude modulation

Transmitter modulation depth can be determined by spectral analysis as well as from the oscilloscope display, however the spectrum analyser will generally permit observation at much higher frequencies. **Fig 3.12** shows a typical spectrum analyser plot for an amplitude modulated carrier modulated at 1kHz.

The double sideband structure is immediately obvious. Note that, unusually, the vertical axis is in linear mode at V/div. The percentage modulation depth 'D' is given by:

D = 100(Vlsb+Vusb)/Vc

For the case shown, the modulation depth is 86%. For low levels of modulation it is better to use a logarithmic display, but the calculation becomes more complex.

When the value of each sideband reaches half the carrier level, 100% modulation depth has been reached. Proceeding beyond this, distortion begins to occur with the number of additional sidebands rising as the distortion level worsens. **Fig 3.13** shows a typical display - obviously this is a situation to be avoided.

Measurement of Frequency modulation

Theory

The Bessel function curves of **Fig 3.14** show the complex relationship between carrier

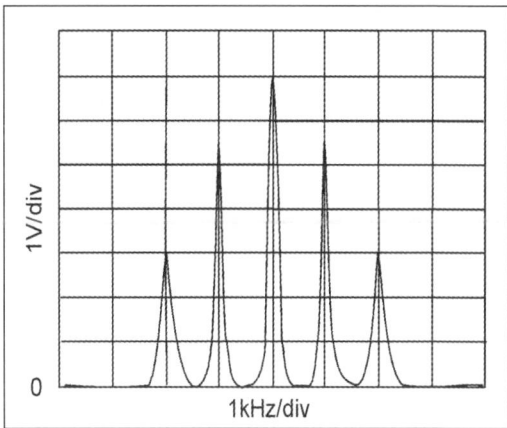

Fig 3.13 Additional sidebands caused by over-modulation

and sideband amplitudes of an FM wave as a function of modulation index 'm'. Observe that the carrier component J0, and the various sidebands JN, all go to zero amplitude (cross the x-axis) at specific values of m.

Looking at the signal in the frequency domain, the associated sidebands will be spaced about the carrier at the modulation frequency. However, for measurement and calculation purposes, 'significant' sidebands are only those which have a voltage of at least 1% (-40dB) of that of the unmodulated carrier – hence the higher order sidebands can be ignored for practical purposes. **Fig 3.15** shows a typical spectrum, and corresponds to a modulation index of 1.2. Clearly only the first two sidebands are 'significant' and of interest here.

Measurement of Deviation

From the foregoing, the carrier and first order sidebands independently go to zero at cer-

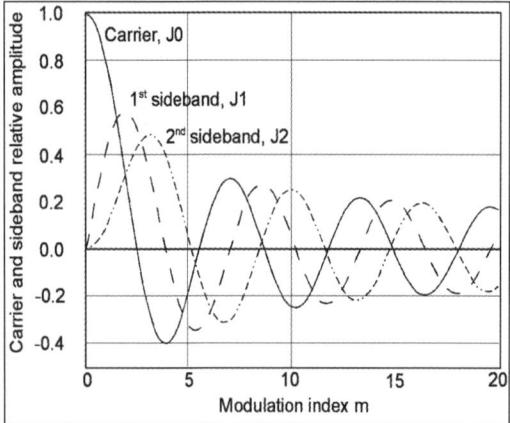

Fig 3.14 Bessel function curves showing relationship of carrier and sideband amplitudes

tain values of m (**Fig 3.16**). The following table shows the specific values of m for which this occurs:

Carrier (J$_0$)	1st sideband (J$_1$)
2.40	3.83
5.52	7.02
8.65	10.17

If a tone of modulation frequency (fm) is applied to the transmitter such that the carrier

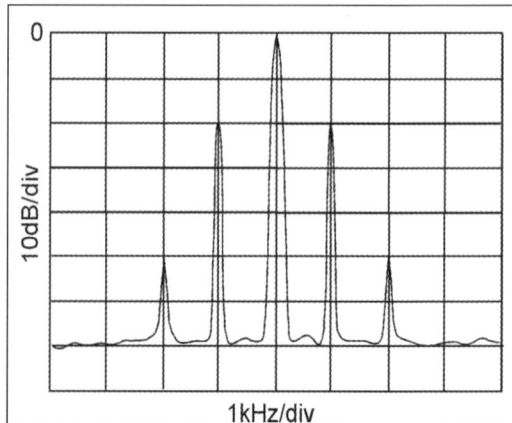

Fig 3.15 Typical narrow band FM spectrum analyser plot

or sidebands are nulled, the value of m is now known. Furthermore, the peak frequency deviation of the carrier (Δf) can be now be calculated since the modulation index (m), is the ratio of carrier frequency deviation to the frequency of the modulating signal, that is:

$$\Delta f = m \times fm$$

This process is called 'the Bessel Zero Method', and can also be used for checking the calibration of deviation meters.

Measurement of Bandwidth

In narrowband FM voice communication, (speech quality) bandwidth is limited and the number of higher order sidebands is restricted, so all sidebands with less than 10% of the carrier amplitude (-20dB) can be ignored. This means that the bandwidth (B) can be calculated approximately by using Carson's Rule, the formula for which is:

$$B = 2(\Delta f + f_m)$$

An example showing both deviation and bandwidth calculations is given in **Table 3.1**.

Electromagnetic Interference & Compatibility (EMI/ EMC) Measurements

EMI: A spectrum analyser with an appropriate transducer attached, such as an antenna or a pick-up coil, is capable of measuring either conducted or radiated EMI. **Fig 3.17** illustrates

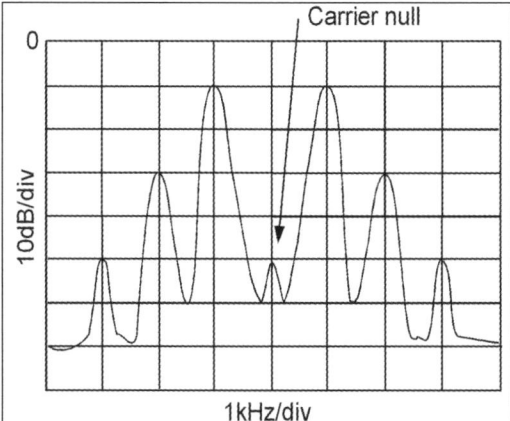

Fig 3.16 Spectrum analyser plot for first carrier null

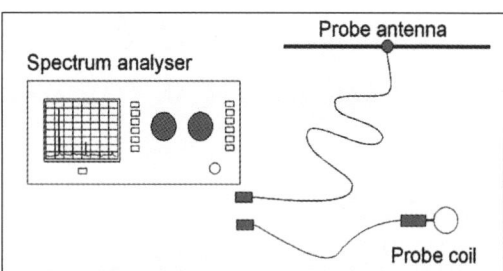

Fig 3.17 Field strength measurement test set-up

a simple set-up used for measuring field strengths.

The antenna is used to convert a radiated field into a voltage for the analyser to measure. The field strength will be the analyser reading modified by the antenna correction factor, as the antenna may not have a flat frequency response.

EMC: This is an important factor for high-frequency circuits in close proximity to each other. For example, in a multi-stage circuit, parasitic oscillation from one stage can couple to a nearby stage and cause unpredictable behaviour. A common technique used to search for spurious radiation is with an inductive-loop probe. The loop probe is simply a few turns of wire that attaches to the spectrum analyser with a flexible coaxial cable (**Fig 3.17** above). Various parts of the circuit can then be 'probed' to identify the location, as

well as the frequency and relative amplitude of an unwanted signal. Once the unwanted signal has been identified, shielding and/or design techniques can be used to reduce or eliminate the cause of the interference.

On a wider point, many Radio Amateurs suffer from RFI caused by modern electronic devices. A particular problem can be interference from wind farms, solar PVs, and VDSL. A useful discussion of how the amateur bands are affected by these emissions, and what might be done to eliminate, or at least mitigate their effects, is contained in the following article:

→ Radio Frequency interference', John Rogers, M0JAV, *RadCom*, March 2017

Signal Tracing & Spectrum Analyser Probes

A really useful feature of the spectrum analyser is the ability to identify multiple signals (ie different frequencies) and to follow the one of interest around the circuit. A simple E- or H-Field probe can be made for this purpose – see Chapter 8.1 'Probes'.

Example

A signal generator or transmitter is adjusted to obtain the first carrier null of the FM signal and a plot similar to the **Fig 3.16** will occur (for practical reasons the carrier may not disappear completely). This first null appears at a modulation index of 2.4 and the sidebands are at 1kHz spacing (f_m).

The peak deviation Δf is therefore 2.4 x 1kHz = 2.4kHz. Using the formula to calculate B, this results in an approximate bandwidth of (2 x 2.4kHz) + (2 x 1kHz) = 6.8kHz.

A similar analysis can be undertaken by finding other nulls for the carrier or sidebands.

The situation is slightly more complicated with wideband FM signals, but there are good tutorials on the Internet.

Table 3.1 Example of FM deviation and bandwidth calculation

4 Equipment for Signal Generation and Frequency Measurement

4.1 Sources

WHETHER TESTING, aligning, calibrating, or fault-finding, an adjustable (but stable) signal source of known frequency and amplitude is often required. Generally, the better the instrument, the more accurate the frequency and output level that can be set, and the more stable (low drift) and cleaner (low phase noise) the signal to be injected will be. It is usual to distinguish between sources in the audio frequency (AF) range, and those working at radio frequencies (RF).

AF Signal Generators

These are useful for testing audio amplifiers, and for modulating transmitters to check their performance. Commercial units covering the audio spectrum in ranges, offering different types of waveform, and providing known levels of output are available (**Pic 4.1**), although a simple AF generator intended for a specific purpose is easily made.

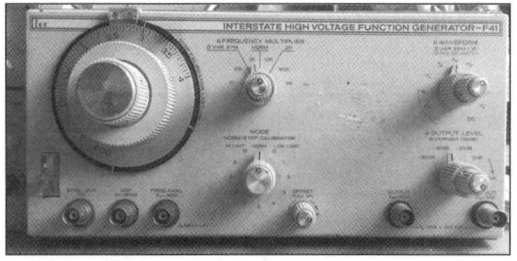

Pic 4.1 An Interstate AF function generator type F41

→ Chapter 8.2 offers a design which provides a sinewave output across the entire audio range, and another which produces a square-wave for use with digital (TTL and CMOS) circuits.

When testing sideband transmitters for linearity and intermodulation products, an audio source consisting of two fixed and non-harmonically related frequencies is required. This can be achieved by using two independent audio sources or by constructing a 2-tone oscillator.

→ Care! Many transmitters employ output devices whose dissipation ratings are exceeded if the full peak envelope power is repetitively produced. Chapter 8.2 describes construction of a 2-tone oscillator which can be operated in burst mode - this pulses the audio sources, such that the mean dissipation is reduced, yet still allowing the performance at peak RF output to be observed.

RF Signal Generators

In a good signal generator, both frequency and output amplitude will be accurately known. There are two basic types:

Free running

Based on a Variable Frequency Oscillator (VFO), they have the advantage that the signal produced is very clean and generally has less phase noise on it than synthesised types. Despite any thermal compensation, their frequency will tend to drift slightly and they need time to stabilise before use. They may incorporate an external socket for a quartz crystal, or some internal quartz crystals, such that the VFO can be switched-out and a crystal switched-in to provide a signal (and its harmonics) whose frequency is known very precisely. A UHF or microwave free-running signal generator may use a tuneable cavity instead as a 'lumped' tuned-circuit.

A handheld GDO may be used as a crude RF signal generator, provided that it is placed remotely from the receiver under test as its frequency is easily pulled with changes in coupling, and bearing in mind that the frequency displayed is only approximate and may drift significantly. Furthermore, the output may have significant harmonic power, and so, as with any unscreened RF source, there is the possibility of interference with local domestic receivers, including televisions.

Synthesised

Virtually all modern RF signal generators use frequency synthesis techniques. This enables frequencies to be entered directly from a keypad or via remote control and allows the output signal frequency to be set very accurately. How accurate depends upon its internal reference oscillator, or possibly an external frequency reference to which it can be locked. The latter facility permits enhanced accuracy and frequency stability with temperature (see Chap.4.3).

Cheaper RF signal generators tend to be based on high stability LC oscillators (**Pic 4.2**). Synthesised ones may cost more (**Pic 4.3**).

Selection

Apart from frequency range and power output range, important factors to consider include:

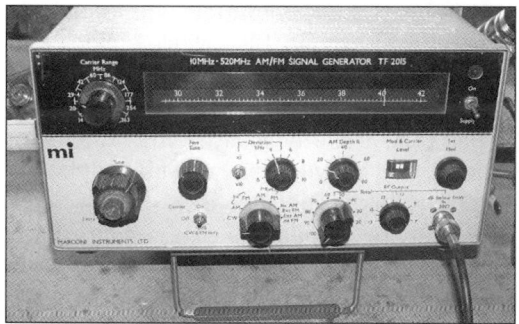

Pic 4.2 An RF signal generator based on an LC oscillator

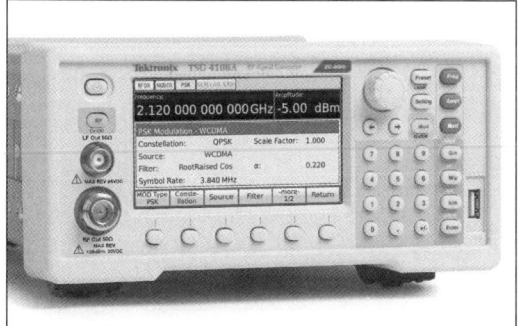

Pic 4.3 A Synthesised RF signal generator
Selection

* Unwanted harmonic and spurious output: all signal generators produce some level of spurious signals. Harmonics are generally much higher as considerable effort is spent in reducing intermodulation and other non-harmonically related spurious signals.

* Accuracy (short and long term): the accuracy of an RF signal generator is important, both in initial setting, and in drift from that setting. Accuracy measurements are usually specified in parts per million (PPM).

* Modulation types supported: RF signal generators usually have AM and FM and some have phase modulation (related to FM). Unmodulated carrier (CW) operation is always available.

* Modulation input: most generators have an internal modulation frequency available (typically 400Hz and/or 1kHz) plus the facility to input external modulation if required.

* Build quality: a poorly screened internal oscillator will emit sufficient RF energy to by-pass internal attenuators and prevent low-microvolt output levels from being attained; therefore the quality of the internal attenuator and the effectiveness of its screening are all-important.

Commercial RF signal generators normally expect to deliver their output into 50Ω, so if a different load impedance is used, a matching pad becomes necessary. However, for simple diagnostic injection, attachment of a common probe with BNC connector may be all that is required.

Fixed Frequency Sources

The crystal oscillator and the harmonics that it produces can be extremely useful in providing a signal, or a series of fixed-frequency signals, whose frequency is known quite precisely. For example, a 1MHz crystal will oscillate at its fundamental frequency, but will also generate a second harmonic at the crystal's first overtone (2MHz); a third harmonic at its second overtone (3MHz), and so on. This combline of harmonics at regularly spaced intervals can be fed into a receiver to check whether the tuning dial is reading accurately: Tuning to the bottom of say, the 40m band (7MHz) the dial should be exactly 'zero-beat' with the crystal's 7th harmonic. If it isn't, the local oscillator of the receiver needs to be adjusted to make it coincide. This is the principle of the 'crystal calibrator', which is sometimes built into older communications receivers because they use a free-running VFO, which is prone to inaccuracy and drift compared to modern units which use a synthesizer. The latter creates local oscillator signal locked to a combline of tiny steps derived from a crystal reference, hence a much greater setting accuracy and frequency stability can be achieved.

A crystal oscillator may also be used in conjunction with a multiplier, (which uses a diode, transistor, or other non-linear device) to increase the number of harmonics, such that a higher-order harmonic can be selected and amplified, perhaps for use a calibration signal, or as a local oscillator in a microwave transverter.

Note that any change of the fundamental frequency of a crystal oscillator, induced by changes in temperature, aging, or circuit loading, will be multiplied by the order of the harmonic selected – so the 10th harmonic will drift at ten times the rate of the fundamental. Using a higher fundamental gives the advantage of reducing this effect. A crystal oven (which increases building costs and power consumption) may be used to keep the crystal at a constant temperature where even a small drift cannot be tolerated.

Frequency sources can be compared to, or locked to, a recognised frequency standard for much greater accuracy and stability – see Chapter 4.3.

→ Chapter 8.3 offers some constructional ideas for RF sources

4.2 Frequency Counters

A frequency counter is an invaluable instrument for alignment, test, and calibration purposes. Among other things, it allows a transmitter frequency to be checked and verified

as in-band; and a receiver to be aligned so that the dial read-out is correct.

The cost of frequency counters varies considerably. For example, £150 or so will buy a 3GHz hand-held unit with a short whip aerial which by which to pick up or 'sniff' for a local RF field; it may also decode and display any CTSS tone modulating the signal. Although workbench units can be much more expensive it is possible to buy something for less (including kits), but operating frequency range, resolution, and accuracy are likely to be compromised. **Pic 4.4** shows a commercial frequency counter capable of counting to 20GHz.

Operation

Fig 4.1 shows a block diagram of a basic frequency counter. In essence, the counter counts the number of input cycles for a given unit of time and displays the result in Hz (or its multiples: kHz, MHz etc). To achieve this, a wave-shaping circuit takes the input signal, amplifies it, and converts it into a rectangular waveform of sufficient magnitude to operate the counting circuits. Meanwhile a clock produces a series of pulses which determine the counting period of the frequency counter. These pulses are typically 10ms, 100ms, or 1s long, and derived from a crystal-controlled clock-oscillator. They are thus of high accuracy, and are applied to a gate, which can be considered as an on/off switch operated by the clock. When the clock-pulse opens the gate, a train of pulses from the wave-shaping circuit is sent to the counting circuit which counts the number of pulses for the duration that the gate is open. This count is then

Pic 4.4 A Marconi 2440, 20GHz frequency counter

frozen, decoded, and used to drive a digital readout, perhaps in the form of an LED or LCD display.

More expensive frequency counters will allow measurement of the period, usually in microseconds or milliseconds. This is useful for frequencies less than about 100Hz and will generally give a more accurate result, though you will need to calculate the frequency (f), which is the reciprocal of the period (T). That is: (f=1/T).

Accuracy and Resolution

The accuracy of a frequency counter clearly depends on the accuracy of the clock signal. It can be increased if the clock oscillator is housed in a thermostatically controlled crystal oven and compared with a standard frequency source for calibration. The resolution of a frequency counter, which is the smallest digit that it displays, is typically 1Hz. However this will change according to frequency

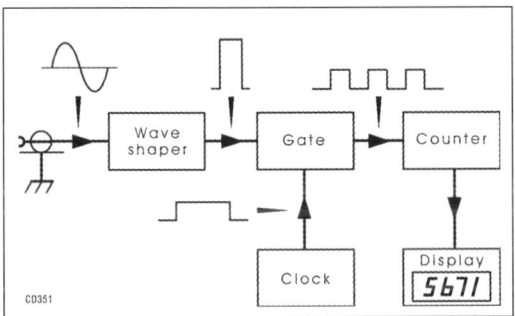

Fig 4.1 Simplified block diagram of a digital frequency counter

range and the length of the clock signal.

Pre-scalers

To increase the range of a frequency counter it is common to put ahead of the input some components that perform pre-scaling.

Fig 4.2 shows that this consists of an amplifier, a wave-shaping circuit, and a high-speed frequency divider (the pre-scaler).

Pre-scaling will divide the input frequency by a known amount (eg 2, 10, 64, or 100) with the resulting signal being applied to the basic

frequency counter (**Fig 4.3**). It should be borne in mind however, that pre-scaling reduces the resolution, so that whereas a frequency counter without pre-scaling may measure to 1 Hz, if a pre-scale of 10 is introduced, then the same counter will only read to 10Hz.

Making measurements

There is usually a high (1MΩ) impedance, and a low (50Ω) impedance input for the lower and higher frequency ranges respectively. The input can be taken from a test probe, direct pick-up off-air using an antenna, or from an RF sampler or coupler.

• Know the maximum input ratings of your frequency counter, and take care not to exceed them

• Keep RF input levels low (only a few tens of mV of RF is needed).

• Consider whether any AC or DC offset voltage is safe. Use AC coupling, especially in higher-voltage valve circuits. Add an external DC block if necessary.

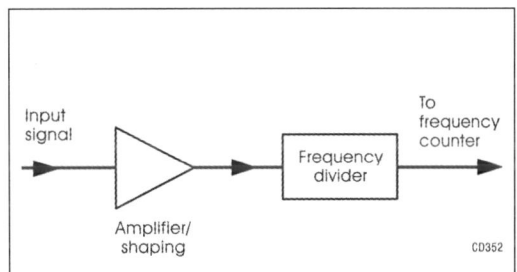

Fig 4.2 Pre-scaling

When making measurements of transmitter frequency, the carrier should be unmodulated to avoid confusing the counter; hence:

• With AM and FM transmitters, do not speak into the microphone or provide any other form of modulation.

• With SSB sets, the CW mode should be used instead, with key down. Take care not to exceed the thermal dissipation limit of the PA.

• With digital transmissions, no superimposed data should be transmitted, just a carrier.

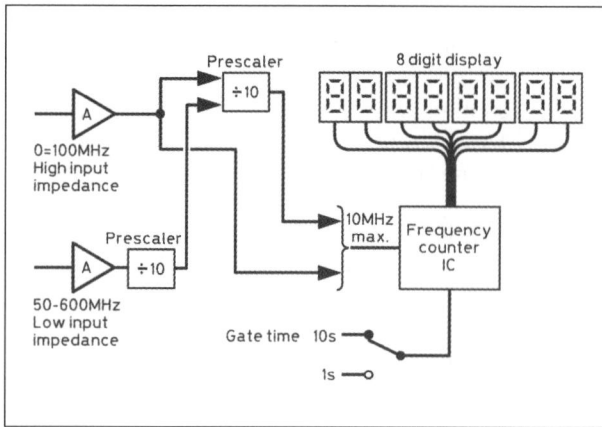

Fig 4.3 Block diagram of a frequency counter with pre-scaling

Probes

Be aware that while a high-impedance probe generally helps to minimise loading of the counter on the circuit under test, if connected directly to the tuned circuit in an oscillator, it may well change the frequency of operation of the circuit by 'pulling' of the oscillator, and may stop it working altogether. This is because the probe can present a shunt capacitance in the order of 10pF at 144MHz, which is a reactance of only 110Ω. The probe may also affect the DC operating conditions. Therefore:

• Do not attach the probe to frequency-determining elements.

• Measure the oscillator frequency after the buffer amplifier.

• Put a 1kΩ resistor in series with the probe.

Another alternative is to use a pick-up loop, as shown in **Fig 4.4**. Ensure only loose coupling, as this minimises influence on the circuit and damage to the counter by input power overload.

4.3 Frequency Standards

When using any equipment that displays frequency or time, the question arises: 'how accurate is it? So for example, when I set 7.000MHz

on a mechanical dial, or digital display of a communications receiver, is it actually listening on that frequency, or is it listening some distance away, perhaps on 7.035? How do I know, and does it matter? For general listening it may well not matter but if looking for a weak CW station on a specific frequency, it may well be missed altogether. The process of removing this error, is known as 'calibration', and requires either a signal of precisely known frequency to be received, or access to a signal generator whose own frequency is traceable to a frequency standard, so that in this example the local oscillator can be tweaked and the error reduced to practically zero.

In the past, the best frequency standard available to most Amateurs was a crystal oscillator, an internally provided or external home-made unit which would emit a signal, perhaps at harmonics of 1MHz and/or 100KHz to mark the amateur-band edges. This 'crystal calibrator' might even be enclosed in its own little oven so as to hold it at a constant temperature to minimise frequency drift. On the receiver there was often a tuning adjustment so that the receive frequency could be shifted a little until the signal from the crystal calibrator peaked at the required dial marking.

Going a stage further, by listening to a standard frequency service transmission (such as WWV), the calibrator itself could be checked against a signal of very precisely known frequency, and carefully adjusted to be zero-beat with it, creating a local transfer-standard, whose frequency is known very precisely.

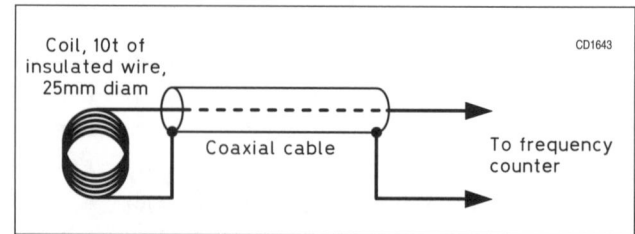

Fig 4.4 Pick-up loop

In times past there were local frequency-measuring competitions on the air (160m and 80m), in which a club station would transmit for a short while and invite listeners to measure the frequency before giving the correct answer. A useful and enjoyable community exercise!

In recent years, standard frequency service transmissions have developed more towards the provision of timing and other information superimposed on the carrier. Although some can still be used for frequency measurement, it is now more difficult to process these transmissions and derive reference signals from them.

As an alternative, it may also be possible to acquire a Rubidium clock salvaged from commercial telecommunications equipment to use as an extremely accurate local standard and derive a reference signal. That said, increasing use is made of the extremely accurate GPS clock transmitted by constellations of satellites and available free to everyone, which can readily provide frequency reference and control signals.

There are thus three types of accurate third-party frequency source available:

1) Standard Frequency Transmissions

In the UK, two sources whose frequency is precisely controlled are the BBC Droitwich Radio 4 198kHz 500kW transmission (to 2 parts in 10^{11} from a Rubidium clock), and the MSF 60KHz 17kW transmission from Anthorn, Cumbria (to 2 parts in 10^{12} from a Caesium clock). However broadcast amplitude modulation on the former, and time-code modulation on the latter can make using them as a frequency reference difficult, although some commercial equipment does exist which can process these signals and output an external reference for driving test and communications equipment. In North America, WWV on 2.5, 5, 10, 15 and 20 MHz and WWVB on 60KHz transmit with a frequency uncertainty of better than a few parts in 10^{13}. The VLF transmissions from MSF or WWVB are preferred for highest accuracy as (unlike their higher frequency counterparts) they do not suffer from significant multipath effects. For more information see:

→ U.S. Dept. of Commerce, National Institute of Standards and Technology (NIST), Time and Frequency Services (especially WWV, WWVB, WWVH).
https://www.nist.gov/pml/time-and-frequency-division/time-services

→ HF Time & Frequency Standard Stations
www.smeter.net/stations/hf-time-frequency.php

→ UK National Physical Laboratory, Time Products & Services; (inc. MSF, Droitwich and GPS Bulletins)
http://www.npl.co.uk/science-technology/time-frequency/products-and-services/time/

2) Atomic clocks

The caesium atomic clock, is based on the natural resonant frequency of the caesium atom at 9.192631770 GHz. This is the primary global frequency standard, with a stability of 1 part in 10^{15}. A rubidium clock is a less accurate, but much more affordable secondary standard. Its frequency is 6.834682610904324GHz and is generated by the hyperfine transition of electrons in Rubidium-87 atoms (Rb87). The amount of light from a rubidium discharge lamp that reaches a photo-detector through a resonance cell drops by about 0.1% when the rubidium vapour in the resonance cell is exposed to microwave power near the transition frequency. A quartz crystal oscillator can thus be controlled and stabilised to the rubidium transition frequency by detecting this light dip. While quartz crystal oscillators have no definite 'wear out' period, a rubidium standard has a 'lifetime' associated with its lamp. During operation, the rubidium within the lamp is gradually used up, and eventually there is too little vapour available for the atomic resonance to be detected and the unit fails. For this reason, amateurs using a rubidium standard choose not to leave them running continuously. Many come on to the surplus market as they have been removed from telephony applications such as cell base stations. They are replaced well before they fail and therefore have a reasonable amount of life remaining. Typical of these is the LPRO-101 which provides a 10MHz output at about +7dBm. It takes some three to five minutes to warm up and lock and gives accuracies generally better than an oven-based quartz oscillator, with ultimate accuracy being achieved after a warm-up period of around 30 minutes.

If a GPS standard is available, this can compared against the rubidium standard to further improve the latter's accuracy.

3) GPS derived frequency standards

Remarkably, there is a really accurate clock available free from space in the form of the satellites delivering GPS. All that is required is a method to receive and decode the signal to produce a usable output. GPS modules which extract and output the 1PPS signal and in some cases a 10kHz signal, can often be found on the surplus market. The GPS signal can be used to control the oscillator in a piece of equipment, creating a highly accurate and stable 'GPS disciplined oscillator' (GPSDO) as a master reference. Brand new GPSDOs (whose frequency can be programmed) can be purchased for around £150 or less. For further inspiration, see:

→ 'A High Precision 10MHz Frequency Standard', *RadCom Handbook* 13th Edn, Chapter 11.19.

A project for a 10MHz reference with stability better than 4 parts in 10^{10} against which other equipment can be calibrated. It simply uses a VCXO module, a GPS counter, and an A/D converter.

→ Adaption of commercial GPS-locked 10MHz frequency standards obtainable from the second-hand market (eg HP Z3801A/ Z3816A):

http://www.realhamradio.com/GPS_Frequency_ Standard.htm

→ For an interesting article covering construction of 10MHz crystal calibrator, an OCXO version, a 10 and 100 frequency-divider (for 1MHz and 100KHz sub-multiple outputs), and aligning the calibrator to a

GPSDO see: 'Homebrew', Eamon Skelton EI9GQ, *RadCom*, December 2016.

4.4 Absorption and Heterodyne Wavemeters

An absorption wavemeter is usually a hand-held unit which absorbs RF energy from a source close by and uses this to drive some form of indicator such as a moving coil meter (**Pic 4.5**). Generally consisting of little more than a diode and a tuned circuit, they are quite simple in design, can be easily made (Chapter 8.11), and require no DC supply. A commercial GDO can usually be used as a wavemeter simply by using the instrument in absorption mode. An earpiece socket may be provided, so that AM audio can be demodulated and checked for quality. Two important uses of an absorption wavemeter are:

• Checking that the correct harmonic is selected for driving the next stage in circuits such as multipliers;

• Checking for the presence of harmonics in the output of a transmitter stage. An absorption wavemeter provides a course, but

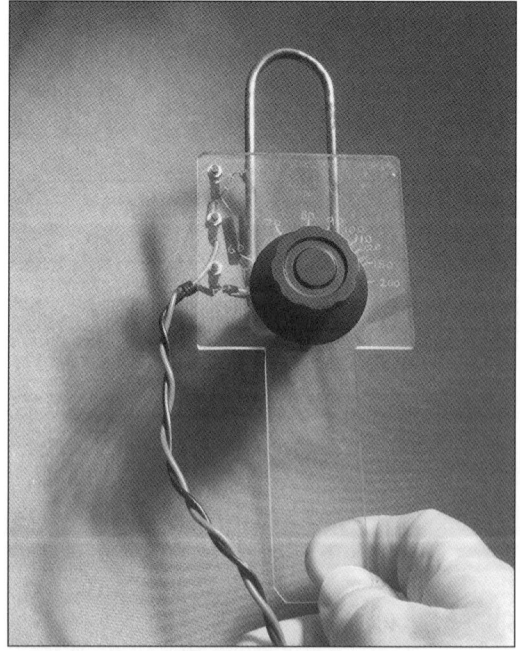

Pic 4.5 A home-made VHF absorption wavemeter

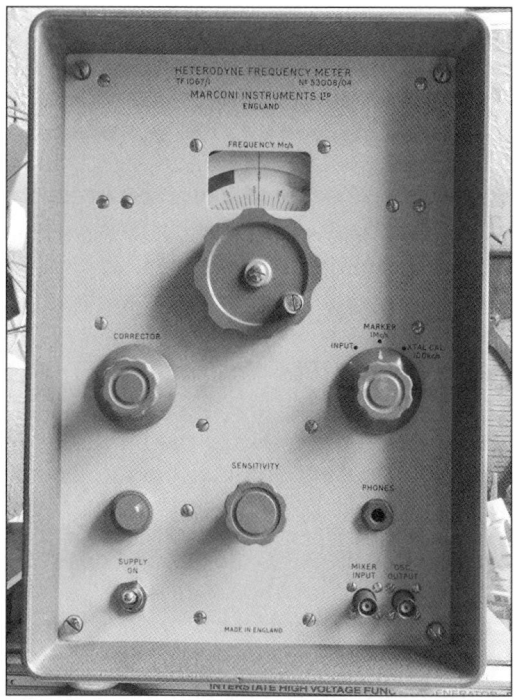

Pic 4.6 A UHF heterodyne wavemeter

unambiguous identification of harmonic frequency, and a rough idea of relative amplitude.

Operation

A good absorption wavemeter should have a continuous frequency range of at least 2:1.

For example, a wavemeter might be tuneable between 60 and 150MHz, which would allow identification of the fundamental output of a 4m transmitter on 70MHz, and assessment of the level of its 2nd harmonic radiating at 140MHz. The hand-held unit is placed within a few feet or a few inches of the source depending on the power output, and the tuning-knob adjusted for maximum indication. Tuning can be quite sharp, and care must be taken not to miss the frequency of interest or overdrive the meter when it's located.

A heterodyne wavemeter is basically an absorption wavemeter, but with a very accurate signal generator attached. The absorption wavemeter picks up or 'sniffs' the transmitted signal, and the signal generator is adjusted until it is on the same frequency or 'zero-beat'. The equipment often has an internal crystal calibrator and look-up tables attached so that the received frequency can be better interpreted from the dial. Heterodyne wavemeters, rapidly fell out of use with the widespread availability of frequency counters, and with the advent of synthesized transceivers having digital readouts. However, at UHF and microwave, old cavity-tuned instruments can still be put to good use. **Pic 4.6** shows such a unit.

5 Equipment for Antenna and Transmission Line

5.1 Antenna System Analysers

THESE CAN MEASURE various features of both antennas and transmission lines, either separately, or when connected together as an antenna and feed system. Typical parameters that can be measured are:

- Impedance (resistance and reactance)
- VSWR (or return loss)
- Resonant frequency
- Bandwidth
- Velocity factor
- Cable loss
- Cable length
- Distance to fault

Measurements are usually undertaken to see how well the antenna or feed-point impedance compares to a system value, often 50Ω. Improving the 'match', improves forward power transfer to the antenna, and reduces reflected power.

Antenna analysis is a specific application of the scalar or vector network analyser (VNA). The former just measures the magnitude of the reflection coefficient while the latter also measures its phase. So while a scalar analyser can measure impedance, the vector network analyser can go a step further and display the values of the resistive and reactive elements that make up that 'complex' impedance.

Such equipment can also be used to check input and output impedances of filters, amplifiers, matching networks, resonance of traps, attenuation, stub performance etc. Simple analysers may just give a readout, whereas those with a PC interface can generally provide Smith-charts plots of complex impedance, phase, etc.

→ For an excellent introduction to using the Smith Chart and practical examples see:

'Antennas', Mike Parkin, G0JMI, *RadCom*, Pt1 & Pt2, March & April 2018.

→ See also Appendix B for comments under transmission line theory and reference to electronic Smith Chart calculators.

→ For an excellent article on using an antenna analyser to measure the properties of cables for feeding and matching purposes, see: 'Properties of Open and Shorted Feed Lines', Roger Paskvan, WA0IUJ, *RadCom*, January 2016

Commercial Products

Professional analysers from the likes of Hewlett Packard, Rohde and Schwarz, Anritsu, Agilent, Boonton etc. may be prohibitively expensive, but are of high quality and designed for everyday use. A typical unit is shown in **Fig 5.1**.

More affordable proprietary equipment designed to cover the amateur bands up to 470MHz is available for under £500. Be aware however, that most inexpensive antenna analysers use broadband voltage detectors because of cost, and these pose some inherent accuracy problems. The main difficulty is unwanted interference from

Fig 5.1 The Anritsu SiteMaster S331D

strong external RF sources - typical would be a very strong signal received on an antenna system. Also the impedance of various connectors, adapters, and lead lengths will affect accuracy, as will internal components such as the diode detector(s) and A to D converters. It is best to note what the accompanying manual says about these problems and how to minimise them. Remember also, that the equipment is probably designed only for a 50Ω system. A popular unit for the amateur market is shown in **Fig 5.2**.

→ Antenna Analysers and VNAs can be expensive, so check-out internet review sites to see how those that have already bought one have fared with their purchase. A really good comparison site is eHam.net at:

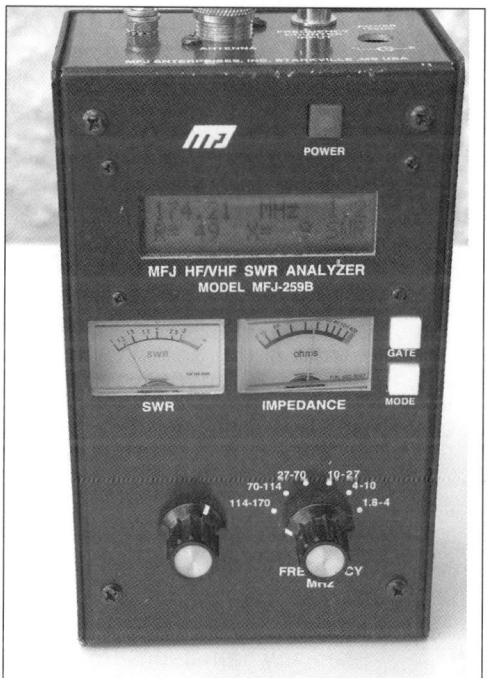

Fig 5.2 The MFJ-259B 'SWR Analyzer'

https://www.eham.net/reviews/products/31

→ See also W8WWV - 'Analyzing Three Antenna Analyzers':

http://www.seed-solutions.com/gregordy/Amateur%20Radio/Experimentation/EvalAnalyzers.htm#Autek%20RF-1

→ Manufactures with products include:

AEA Technology:

http://www.aeatechnology.com/via-echo-vector-network-analyzers

Array solutions:

https://www.arraysolutions.com/antenna-analyzers

Autekresearch:

http://www.autekresearch.com/

MFJ Enterprises:

http://www.mfjenterprises.com/Categories.php?sub=0&ref=7

Mini Radio Solutions (MRS):

http://miniradiosolutions.com/

Rigexpert:

www.mixw.co.uk

Times Technology:

http://www.timestechnology.com.hk./

Universal Radio:

https://www.universal-radio.com/catalog/meters.html

→ You can also build your own from a kit - examine the following websites:

VK5JST (Apparently, over 17,000 have built; note the excellent comments on this build by M5POO):

http://users.on.net/~endsodds/aamk7.htm

http://www.m5poo.co.uk/vk5jst-aerial-analyser/

N2PK

http://n2pk.com/

DG8SAQ and DG5MK (some new products here):

https://sdr-kits.net/DG8SAQ-VNWA-models and FA-VA5 vector antenna analyzer

→ For more on related self-build projects see Chapter 8, sections 9-12.

5.2 Antenna Current Probe

This is a probe which couples into the field surrounding an antenna by clipping over the wire carrying the RF current 'I'. A voltage is induced, given by V = jωMI where M is the mutual inductance, ω= 2πf, and j the operator √-1, which is applied to the diode detector and meter. See **Fig 5.3**(a)

Single-loop type: To calibrate, a small loop of say, 25mm diameter may be inserted in the output line, which will also allow connection to an oscilloscope.

The loop of the probe to be calibrated is coupled into the small loop to obtain a full-scale reading and the power level reduced in steps. In use, the probe is held at a suit-able distance from the transmission line de-pending on the sensitivity required. Usually a rough estimate of current is sufficient and, having settled on a measuring distance (eg 15mm) as judged by the eye, the repeatabil-ity of readings is adequate.

Double-loop type: Some-times, when making obser-vations of an open wire line having high VSWR, read-ings may be affected by ca-pacitance be-tween the user and ground, a typical symptom

Pic 5.1 A two-loop current probe

being dependence of the reading on which way round the probe is held. This effect can be minimised by the balanced diode circuit shown in **Fig 5.3** (b). As a further measure, the meter should be as small as possible - see **Pic 5.1**. A long and insulated handle may also help.

→ For original article see HF Antennas for all Locations, 2nd Edn, Les Moxon, G6XN, RSGB 1993.

→ For more ideas on making such a probe, see 'A clip-on RF current meter', Chapter 8.11.

5.3 Field Strength Meters

These can be used to produce polar plots (radiation patterns) of antennas. That said, Field Strength Meters are often not used by amateurs to make field strength measurements, but rather, just to indicate the presence of RF. Commonly, the transmitter system is tweaked for maximum indication on the meter to maximise the radiat-ed power. Costing just a few pounds, these units usually rely on some form of telescopic an-tenna to pick-up the near-field

Figure

(a)

Line

Square loop

1000pF

μA

0–50μA meter

Antenna or line under test

I

Unwanted voltage pick-up causes adding currents in one half-loop subtracting in the other

(b)

Twist or tape together

0·001 μF   0·001 μF

470Ω   470Ω

μA

Small 0–50μA meter

Fig 5.3 Two forms of current probe

and a simple tuned or un-tuned detector arrangement (**Pic 5.2**). The absorption wavemeter, or a dip oscillator in absorption mode, can be used for a similar purpose.

→ In connection with this, there is an excellent Windows program 'PolarPlot' (free for non-commercial use) by Bob Freeth, G4HFQ, which involves a receiver, an antenna on a rotator, and a PC with a soundcard, which will measure and display the radiation pattern of a beam antenna (**Fig 5.4**).

It relies on an antenna receiving a signal,

Pic 5.2 A typical field strength meter for Amateur use

and the audio output (with AGC disabled) plotted with rotation. The accompanying help files assist configuring the set-up for taking measurements. Download PolarPlot at: *http://www.g4hfq.co.uk/download/PolarPlotSetup.exe*

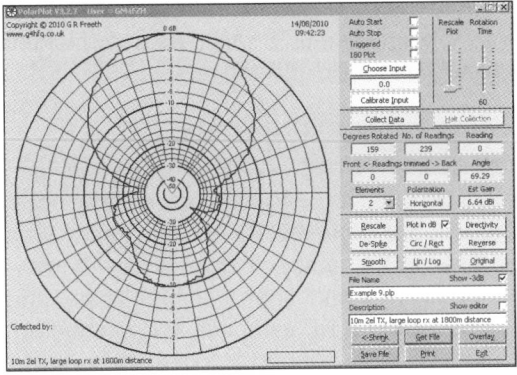

Fig 5.4 Typical display using PolarPlot

→ For a short tutorial on the relationship between field strength and received power, see:

http://www.giangrandi.ch/electronics/anttool/rx-field.shtml

5.4 RF Bridges

Two important types are:

The Noise Bridge

Often used to set-up antennas, it is especially useful as it can determine both the 'real' (resistive) part *and* the 'imaginary' (reactive) part of an impedance. The impedance to be measured is simply connected to the 'unknown' socket, the noise generator switched on, and the receiver tuned to the frequency at which the test is to be made. Two controls (one which varies a resistance and the other a capacitance) are then adjusted to obtain a minimum noise reading on the receiver S-meter. This null occurs when the resistive and reactive parts of the impedance being measured coincide with those inside the bridge, and the values of R and X can be read-off the instrument. Note that the reactance X could be positive (inductive) or negative (capacitive) so the control will be marked accordingly (**Pic 5.3**). The value of L or C at the frequency in use can now be calculated since:

$L=X/\omega$ and $C=1/\omega X$ where $\omega=2\pi f$

The Return Loss Bridge

A return loss bridge (RLB) is usually a wideband resistive bridge network which can be used to measure the impedance of coaxial cables, antennas, tuning stubs, filters, duplexers, cavities etc. It compares an unknown impedance with a known impedance (typically 50Ω) and a voltage is produced which is related to the impedance mismatch. The directivity of the bridge is a measure of the quality of the bridge and must be much better than the return loss that you expect to measure. The directivity and insertion loss of the bridge will vary with frequency.

RLBs are simple to build and have a large

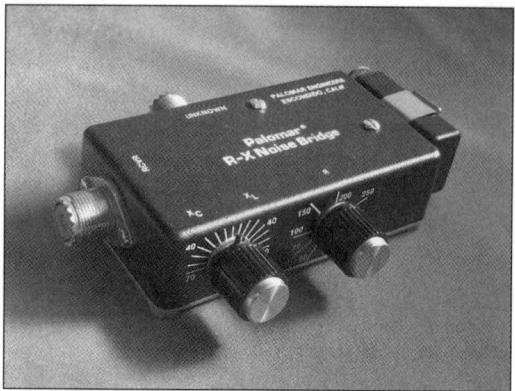

Pic 5.3 A commercial noise bridge for the Amateur market

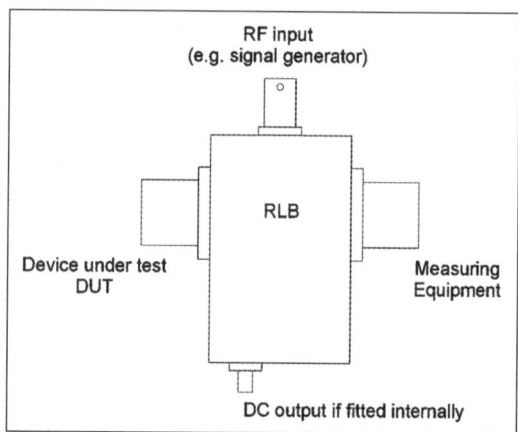

Fig 5.5 Typical connections when using a return loss bridge

bandwidth, typically 1 to 500MHz – somewhat wider than a typical VSWR meter. **Fig 5.5** shows a typical arrangement for using an RLB.

→ For a few short practical articles on RLBs see;

http://www.qsl.net/n9zia/rlb/index.html

http://www.vk2zay.net/article/179

www.wetterlin.org/sam/Reflection/ManualReturnLoss.pdf

'Design Notes', Andy Talbot, G4JNT, *RadCom*, January, 2018

→ Go to Chapter 8.10 for ideas on how to make an RF bridge or either type.

5.5 Power and VSWR Meters

Power Meters

There are two main types:

Through-line

Based on RF samplers or couplers they sample the power on the transmission (feed) line and indicate a power level. They are usually directional, so that both forward and reverse power can be measured.

Absorption

These incorporate a sensor which absorbs the power, and generates a signal proportional to this power to drive a meter or similar. Sensors involving heating use detectors such as thermistors, thermocouples and the RF (hot-wire) ammeter. They are good at measuring average power in systems because the heating effect produces an averaging, and so they measure power irrespective of waveform. Good for signals such as CW, AM, FM and pulse waveforms they are unable to measure instantaneous power in signals such as SSB.

For instantaneous measurements, a diode-based detector is normally used, and this is the sensor arrangement behind virtually all VSWR and power meters that the Radio Amateur will buy. The Schottky diode has a good frequency response, and can typically be used to measure power levels down to about -20dBm, but then the forward voltage-drop of the diode becomes a problem. With suitable signal processing, the resulting output can be used to display various parameters. When using higher powers, RF samplers and couplers are required as discussed later.

Using Other Instruments

Oscilloscopes and spectrum analysers can also be used to measure power; the oscilloscope has a relatively low frequency limit compared to the spectrum analyser, however the latter cannot cope with high power and so must rely on RF samplers and attenuators to reduce the input power to a suitable level.

Pic 5.4 shows the ubiquitous Bird power meter with its plug-in elements that give different power and frequency ranges.

→ See 'RF & Microwave Power Meter Tutorial - a summary or tutorial description of RF & microwave power meter technology and measurement' by radio-electronics.com at: *http://www.radio-electronics.com/info/t_and_m/rf-microwave-power-meter/basics-tutorial-introduction.php*

→ Ready-made power meters and kits for home-construction are available on the internet.

→ See Chapter 7.1 for transmitter power measurement and 8.12 for power measurement projects.

VSWR Meters

These combine a directional coupler and detector to enable power to be measured in both directions. From this, the reflected power and the VSWR can be deduced and displayed. It is common to have two analogue meters, or one analogue meter with two needles, so

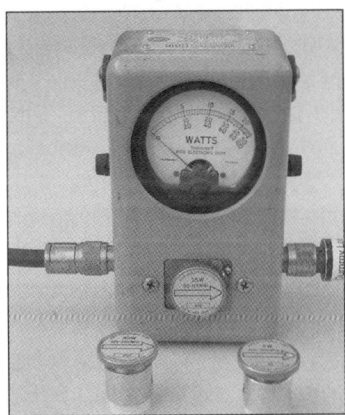

that forward power and VSWR can be displayed simultaneously (**Pic 5.5**). This gives the operator a val-

Pic 5.4 The Bird power meter

uable indication of the power being delivered into the aerial feeder, the quality of the power transfer or 'matching' and a visual indication of any fault condition, such as a short, or an intermittent caused by weather affecting the integrity of the aerial-feeder arrangement.

Fig 5.6 shows the relationship between forward and reflected power and VSWR, which are the parameters of practical interest, while Fig 5.7 demonstrates how reflected power increases dramatically with increasing VSWR. A 1:1 VSWR is ideal, but from a practical viewpoint, it is worth trying to get a SWR better than 2:1 (better than 11% reflected). The guidelines of **Table 5.1** are suggested practical conditions and the actions that should be taken.

→ See Chapter 8.9 for VSWR meter construction projects.

→ See Appendix B for more transmission line theory.

5.6 Attenuators and Matching Pads

Often, a signal cannot be fed directly into a piece of test equipment, such as a frequency

Pic 5.5 A typical commercial power & VSWR meter

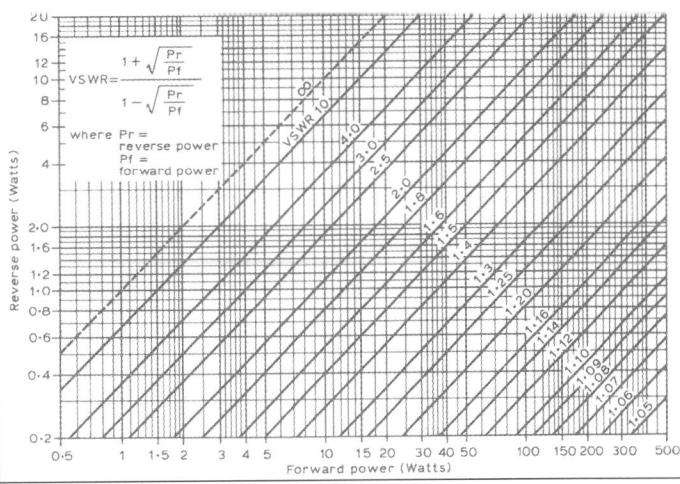

$$VSWR = \frac{1 + \sqrt{\frac{Pr}{Pf}}}{1 - \sqrt{\frac{Pr}{Pf}}}$$

where Pr = reverse power
Pf = forward power

Fig 5.6 Relationship between VSWR and forward/reflected power

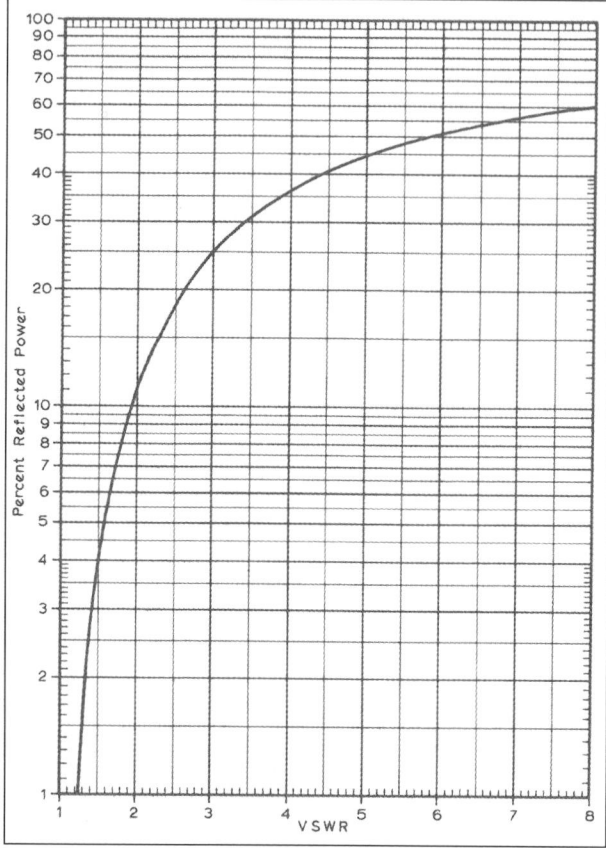

Fig 5.7 Reflected power as a percentage versus SWR

tenuator:

• Fixed attenuators: These have a fixed value (eg 3dB, 10dB, 20dB) and often fit in-line on coaxial cable. Typically terminated in connectors such as N-type, BNC, and SMA, they are primarily for test bench or field use.

• Step attenuators: Larger, and more likely to be used on the test bench, stand-alone types may have a total attenuation of around 100dB, which can be switched in a combination of decade (10dB) and unit (1dB) steps. Step attenuators may also be found incorporated into items of test equipment such as signal generators in order to change output levels.

• Variable attenuators: These are normally used in applications where it is necessary to continuously vary the level of a signal. They are often used where accuracy is not a prime requirement.

Resistors are generally used to provide fixed levels of attenuation. PIN diodes are normally used where a continuously variable level is required.

When making, buying, or using an RF attenuator, some points to consider are:

• Attenuation: This is the primary specification for the attenuator, and is quoted in dB.

• Attenuation accuracy: How accurate is the attenuation, for example ±0.1dB or ±0.5dB? It is often necessary to know the accuracy, especially when equipment is being tested.

counter, spectrum analyser, or oscilloscope, because its level is too high and may damage it.

This requires the signal to be reduced or 'attenuated'. An attenuator of known characteristic is thus used to 'pad down' the incident power by a given amount (eg 30dB or 1/1000th).

There are several types of external RF at-

VSWR	% Reflected Power	Comment
0.0 - 2.5	0-18	solid-state rig VSWR-protection starts to operate; try looking for an improvement at higher SWRs
2.5 - 5.0	18-45	valve equipment will probably survive - look to improve the SWR to nearer 2:1
5 upwards	45-100	check the feed/antenna system - there is a problem!

Table 5.1 Guidelines for various SWRs

- Frequency response: The attenuation is likely to vary with frequency; the higher the frequency the more the error. This is especially likely to be true for homebrew attenuators, and may result from the frequency characteristic of the resistors used, stray inductance and capacitance, or coupling between input and output. Some RF attenuators may have an attenuation-frequency chart supplied with them.

- Impedance: RF attenuators are designed for use in systems with a given characteristic impedance. For amateurs, 50 ohms is probably the most common although 75 ohms is also used. It is possible to obtain RF attenuators with other impedance values but these are not common.

- Power handling: In order to attenuate the signal level, RF attenuators dissipate or absorb the unwanted power. For small signal application, power dissipation is not an issue. For applications where signal levels are higher it is necessary to ensure that the RF attenuator can handle the power levels used. Power capabilities for RF attenuators may be quoted in watts or dBm.

Matching pads enable the matching of different impedance systems (eg 50 ohm and 75 ohm) and typically include attenuation.

→ For constructional tips and how to calculate the elements of different types of attenuator and matching pad, see Chapter 8.6

Commercial products

Good quality fixed attenuators and matching pads can be found on the internet. They may seem expensive, but they are made to have an essentially flat response over a very wide frequency range, such as DC to 2GHz, DC to 18GHz etc. **Pic 5.6** shows a selection of such units. Typical manufacturers are Narda, Hewlett Packard and Weinschel.

Step-attenuators are available, but they are not cheap. **Pic 5.7** shows a unit by Telonic which gives 0-110dB in 1dB steps, DC to 2GHz.

5.7 Samplers & Couplers

Frequently, a test may require a tiny portion of the signal travelling from A to B to be 'siphoned-off', or 'sampled' for monitoring and measurement purposes. Crucially, the aim is to disturb the characteristics of the transmission medium (which is often, but not always, coax) as little as possible in order to avoid unwanted effects such as loss and reflections. The 'proper' way to do this is with a coupler, or directional coupler, which is designed for the task. A coupler has a fixed relationship between through signal and sampled signal, eg 30dB - ie the sampled output is always 1/1000th of the through signal. The range over which this is valid is always specified, eg 10MHz to 500MHz. Beware of power above the quoted range (eg in transmitter harmonics), where coupling may be tighter and sampled output much higher.

Fig 5.8 shows a typical directional coupler; ports 1 and 2 represent the through-line, whilst ports 3 and 4 provide a sample of the forward and reflected voltage. The important parameters are the coupling factor between 'sampled' and though-line (eg -30dB) and the directivity – the fact that ports 3 and 4 show accurately the forward and reverse voltage on the though-line. Note: it is the forward and reverse voltages that are measured not the power, succeeding meters are then scaled in power.

When making, buying, or using a directional coupler, pay particular attention to such specifications as input power limitation, through-line insertion loss, directivity, coupling coefficient and operating frequency.

In some situations, a cheap and simple 'sam-

Pic 5.6 Typical in-line fixed attenuators

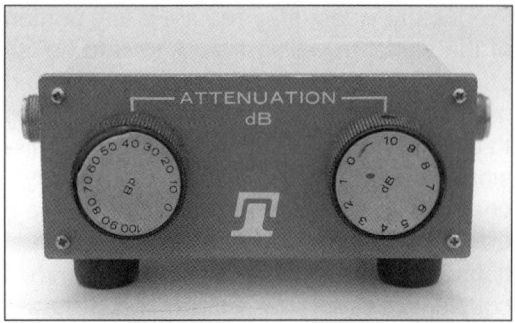

Pic 5.7 A commercial step-attenuator

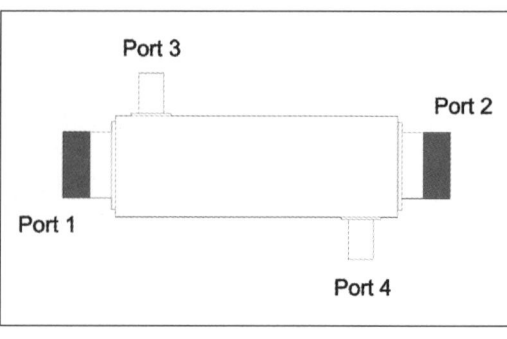

Fig 5.8 Concept of directional couplers

pler', may be quite sufficient, provided that its insertion does not significantly degrade the system under test. The sample may well be of unknown magnitude and vary with frequency, but this may be immaterial if all that is required is to obtain a frequency measurement or monitor a signal waveform. **Fig 5.9** shows the basic block diagram for taking a 'sample' of RF for measurement purposes. A more unfamiliar use of the sampler/coupler is using a signal generator to couple a signal onto (rather than off) the through-line (assuming no transmit power is being passed) via the 'sampled' port.

The following parameters should be considered when using or making these items:

• Impedance: Minimising disturbance to the through-line that is being sampled; ie minimising effects on VSWR.

• Attenuation: Protecting the test equipment from high RF voltage. If transmitter outputs are involved, the power rating of the coupler or sampler must be adequate for the job in hand.

• Bandwidth: This is limited by the type and quality of coupling used, components employed, and the mechanical arrangement. Problems occur as the frequency rises, and as stray capacitance and inductance become more dominant.

• Coupling type: Should the coupling be capacitive, inductive or resistive? The latter implies a physical electrical connection.

• Directivity: High directivity is required for

directional couplers.

The term coupler/detector has been used where components have been included to convert RF to a DC voltage for further processing. This may be required to measure parameters such as forward and return power and VSWR using DC meters.

Commercial Samplers and Couplers

Samplers and couplers can be bought and these will have known frequency characteristics and coupling coefficients. **Pic 5.8** shows a typical sampler and a directional coupler.

→ See Chapters 8.7 & 8.8 for couplers which are easily home-built.

Calibrating RF Couplers

Receiver method: this requires a suitable signal generator with a calibrated output, two 6dB attenuators, a good dummy load and a receiver with an S-meter. Connect the signal generator straight into the receiver and at a suitable frequency set the output level to give

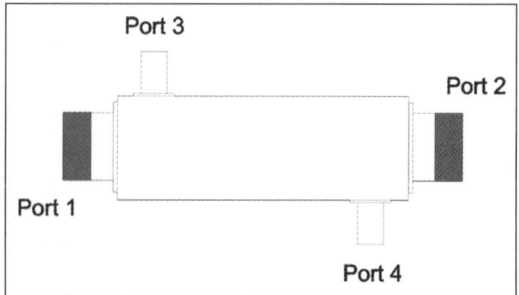

Fig 5.9 Basic block diagram for taking an RF sample

an S9 meter reading – see **Fig 5.10** (a). Note the signal generator output in dB. Now introduce the coupler between the two 6dB attenuators (which serve to stabilize the transmission line impedance and reduce reflections) as indicated in **Fig 5.10** (b). Reset the signal generator to give an S-meter reading of S9, and note the new signal generator output in dB. The difference of the two readings is the coupling factor.

Spectrum analyser method: A signal generator is set to, say 0dBm, run into the spectrum analyser and the level checked. The signal generator is then connected to the through-line of the coupler which is terminated in a good 50Ω dummy load. The sample output is coupled to the spectrum analyser and the difference in level read from the screen. This is the coupling factor.

Care! A 3-port directional coupler (comprising a through-line and a coupled port) will have a load on the forth port which should not be removed. If a 4-port directional coupler is being calibrated, the second sampling port should always be properly terminated. It can be tested with the through path either way round, and the symmetry should be reflected in similar coupled results both ways round.

5.8 Loads

Dummy loads are used by Radio Amateurs both in setting up equipment and also in testing. It should simulate an almost perfect antenna/transmission line system into which the transmitter can run, but unlike an antenna, does not radiate. Points to consider are:

• Impedance: The dummy load should be nominally resistive and have the correct resistance value for the system in use. The majority of the time this will be 50 ohms.

• Power handling: The dummy load should be capable of dissipating the heat produced when running the transmitter into the load. Whilst the dummy load may not be continuously rated it should be capable of coping with the power on a known duty cycle, eg full power for 10 seconds, off for 30 seconds. Low power loads may be needed for couplers etc.

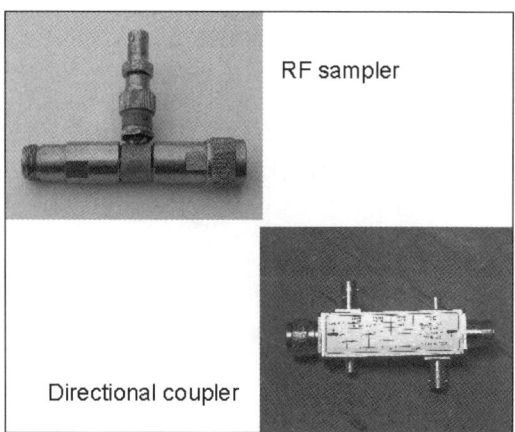

RF sampler

Directional coupler

Pic 5.8 A commercial sampler and directional coupler

• Frequency response: This is an outcome not only of the resistive elements used, but of its physical construction and the connector chosen. The RF dummy load should provide a return loss better than 20dB which equates to an almost 1:1 VSWR within the intended frequency range. A return loss of greater than 30dB would be ideal.

• Radiation leakage: The dummy load should provide good screening and minimise radiation leakage.

As well as resistive dummy loads, open- and short-circuit loads can also be procured. These are designed to look like an open or short over a specified frequency range, and are typically used to set-up a Vector Network Analyser ready for the measurement

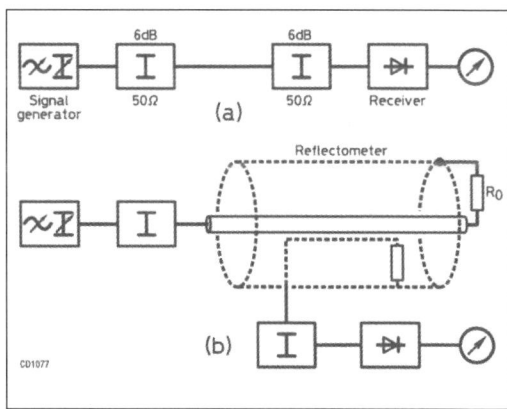

Fig 5.10 Insertion method of measuring coupling

Pic 5.9 Some typical commercial dummy loads

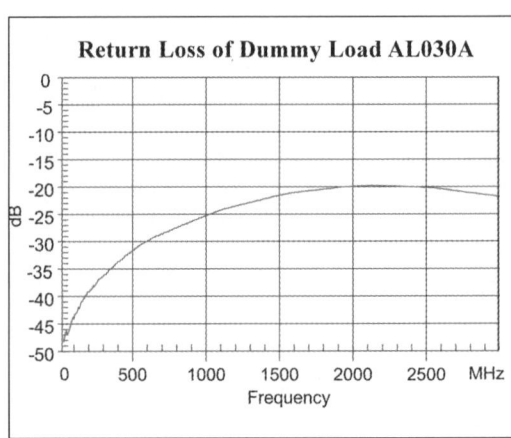

Fig 5.11 Typical return loss of a Celwave AL030A dummy load with frequency

of complex impedance (R±X). They are very precisely made, so that at the port where they are attached 'the reference plane', there is a reflection coefficient of magnitude 1, and a phase of 0/90 degrees. The display (invariably a Smith Chart) can be adjusted accordingly, and the instrument is ready for use.

Commercial dummy loads

Pic 5.9 shows some commercial dummy loads. **Fig 5.11** shows the typical frequency re- sponse of a Celwave AL030A 50Ω 30W dummy load between 25MHz and 3GHz (taken using an Agilent Sitemaster S331D). Always try to obtain such a graph of return-loss against frequency for your particular load, so that you know what its characteristics are at the frequency of interest.

6 Equipment for System Tests & Alignment

6.1 Network Analysers & Wobbulators

Network Analysers

A PROFESSIONAL GRADE network analyser is used daily in industrial laboratories and production lines for the accurate characterisation of a 1- or 2-port network, whether it be perhaps an aerial feed, filter, or an amplifier. As a minimum, it will display parameters such as gain-response over a frequency range, as well as impedance, phase, and VSWR. A rectilinear and (more usefully) a normalised Smith Chart polar display will be available. However, these instruments are often prohibitively expensive for the Radio Amateur, even on the second-hand market. Such instruments are not easy to make owing to their complexity and the accuracy required, although these days it is possible to obtain a kit with calibration accessories for under £500. See the sources below.

→ A 1KHz-1.3GHz 2-port PC-controlled VNA kit, developed by DG8SAQ with software and updates, is available from: *http://www.sdr-kits.net*

→ A battery-portable VNA with Bluetooth for remote monitoring (~£250), can be seen at: *www.nevadaradio.co.uk*

→ Review: 'MetroVNA Pro Touch, a Vector Network Analyser', Steve White, G3ZVW, *RadCom*, August 2016.

→ For simple impedance measurement, a much cheaper alternative is to use a bridge such as described in Chapter 5.4. For 1-port measurements, an antenna analyser may also suffice (see Chapter 5.1).

Wobbulators

When it comes to measuring gain-frequency response (ie how the gain varies over the passband of an amplifier or filter), what is the alternative to a network analyser? While the gain could be spot-measured and plotted at a number of frequencies over the band of interest, a much better solution might be a 'wobbulator'. This can be thought of as a cut-down version of the Network Analyser in gain-response mode. The concept is quite simple: an input signal is swept across the frequency range of interest in sync with a detector, traditionally an oscilloscope, whose trace moves across the screen while monitoring the output of the device under test. This arrangement thus gives a 'picture' of the response. In Amateur circles, it is an extremely useful tool for examining (for example) a receiver's IF response, and for optimising it. A schematic of its operation is given in **Fig 6.1**.

→ Wobbulators do occasionally crop-up on the second-hand market, and in recent years a simple kit based on a couple of chips and a few components has been developed for use with the Raspberry Pi. A Raspberry Pi Wobbulator kit is available from: *http://www.cutpricecables.co.uk/irpiwobbulator.html*

→ M5POO gives useful notes on getting the above kit going at:

http://www.m5poo.co.uk/getting-the-rpi-wobbulator-working-on-a-raspberry-pi-2/

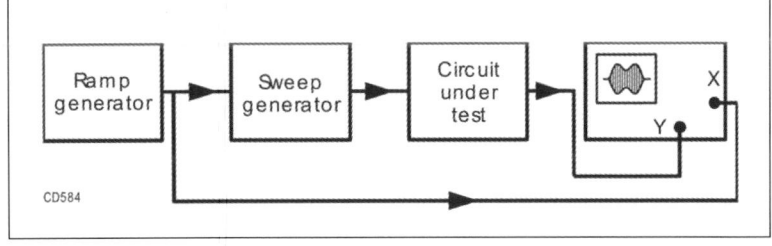

Fig 6.1 Block diagram of an IF passband measurement using a wobbulator

6.2 Noise Figure Meters

This instrument operates by feeding a calibrated noise source, (that is one whose noise noise characteristics are known), into the input of the device under test, and measuring the output. The noise source is automatically switched on and off by the noise figure meter, and the instrument works out the noise figure by comparing the change in signal-to-noise ratio at the input, with that at the output (**Fig 6.2**).

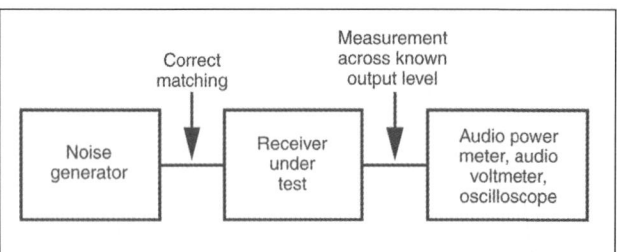

Fig 6.2 Block diagram of a measurement using a noise figure meter

It sounds easy, but while anyone can connect a noise-figure meter to a receiver, mixer, or amplifier, and display a noise-figure in dB, obtaining a valid result through correct measurement technique and the proper application of corrections for factors such as mismatch loss, filter bandwidth, and temperature, requires a level of thought and skill that can be a challenge even to the most professional engineer.

Noise figure meters typically cost at least a thousand pounds, although they do occasionally appear on the second-hand market for less. But beware – the instrument is useless unless it has comes with a noise source (or 'head') of known ENR, and ideally has been recently calibrated (ie validated against a known standard).

At the end of the day, the amateur is usually less interested in numbers and more interested in how well he/she can hear DX. This means buying receiving equipment that has already been verified as having a good noise figure, with the lower the better. When constructing, rather than trying to measure noise figure, a more practical approach is to tune for best signal to noise, or better still, use an uncalibrated noise source to tune for best signal to noise ratio as described in Chapter 7.2.

→ See Chapter 7.2 and Appendix B (Section 1) for more information on noise figure and how to measure it.

6.3 Modulation Analysers

The ability to measure the modulation characteristics of one's own transmission is highly desirable, not only to assess compliance with one's licence conditions, but also to be a good neighbour in crowded band space.

Commercial equipment

A modulation meter will normally measure both AM (modulation depth) and FM (deviation). **Pic 6.1** shows a typical unit, 1980s vintage, covering AM and FM and frequencies from 1.5MHz to 20GHz. They tend to be older items as the communication test set has replaced the modulation meter in the modern commercial setting.

Measurements of an SSB signal are somewhat more difficult, but can be made with an oscilloscope.

→ See Chapters 3 and 7.1 for theory and practical methods of modulation measurement.

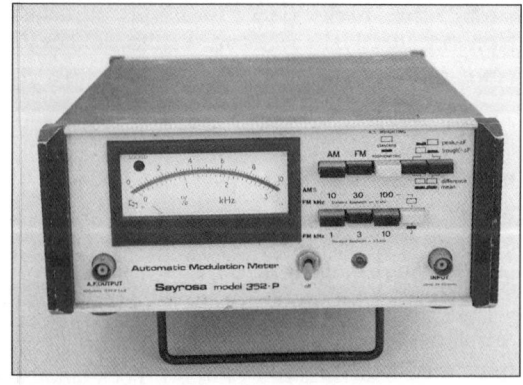

Pic 6.1 An AM/FM modulation meter

6.4 Signal Injectors & Tracers

These are an extremely useful for quickly pinpointing a fault when servicing equipment. The signal injector is typically a 1KHz multivibrator (deliberately rich in harmonics), and the tracer an audio amplifier with a gain adjustable up to perhaps 40dB. The signal injector enables you to work from 'front-to-back' (input to output), and the tracer from 'back-to-front' (output to input). So in a common scenario of nothing coming out of the receiver speaker, you might attach the tracer to the centre and earth of the volume control to see whether any demodulated IF (audio) is present, and if so, follow the signal path through the circuit to the point where it disappears, revealing the area of the fault. Alternatively, if there is no audio at the input, the signal injector can be used to provide it, working from the speaker back towards the input. Some injectors may be broad enough to prove useful injection at low RF and IF so that the signal can be traced through.

A simple signal injector and signal tracer will have no more than half a dozen components each. These are not critical, are cheap, and easily obtainable if not already in your junk box. See the internet for a wide variety of circuits. Commercial units are available (**Pic 6.2**), and a kit can be purchased for around £10.

→ ESR Electronic Components Ltd (authorised dealer for Velleman) *http://www.velleman. co.uk/contents/en-uk/p261.html*

6.5 Power Supplies

Commercial items of test equipment will invariably operate from AC mains or be arranged to operate from batteries. Kits and home-brew in contrast, could be organised to operate from batteries (ideally rechargeable to save on cost) or an AC mains-powered DC power supply. In the latter case, it may sometimes be possible to press into service a cheap commercial wall-adapter, but take care that not only are the voltage and polarity correct, but that it can supply the required current continuously without exceeding its rating, and that the regulation and ripple (which is likely to be poor), is adequate for the task.

When buying power supplies, expect to pay a reasonable price for new or ex-industrial units, as these are likely to be designed for continuous use, and with margin. Cheaper units tend to have a lower value of smoothing capacitor with a higher voltage on them in order to keep the ripple voltage above the minimum required for the regulator. This means that if they fail they may put excessive voltages on equipment and cause additional failures and damage.

Theory

The main issues are transforming the mains to the correct voltage range and then conversion to DC. **Fig 6.3** shows two arrangements for conversion of AC to DC using full wave rectification.

If the secondary voltage of the transformer is Vs (RMS) and the voltage drop across each diode is Vd (=0.6V for silicon), then for **Fig 6.2**(a) the voltage across the smoothing capacitor (Vc) is given by:

$$Vc = \sqrt{2}Vs - 2Vd$$

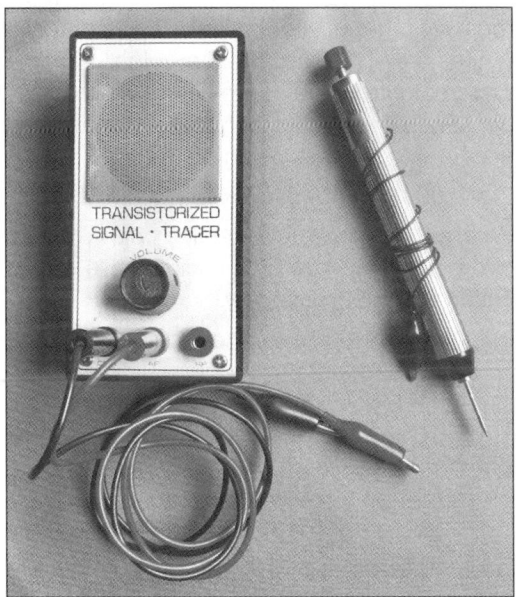

Pic 6.2 A signal injector and signal tracer

For **Fig 6.2**(b) it is given by:

Vc = √2Vs - Vd

This is the voltage with no load. When a load is applied across the capacitor, ripple occurs and the voltage is obtained as shown in **Fig 6.4**.

In addition, the transformer has regulation - ie on heavy loads the output voltage will sag. Providing one obtains a transformer from a reputable supplier, the ratings quot-

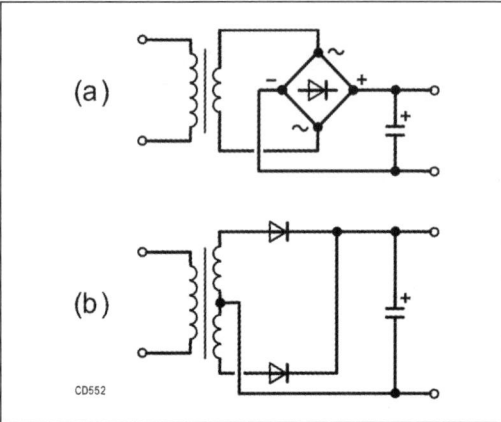

Fig 6.3 Two methods of obtaining full-wave rectification

ed will be when on load, for example a 12V, 5A transformer will have an RMS output voltage of 12V when providing an AC current of 5A. A regulation figure is usually quoted, which for this example may be 7%. This translates to an output voltage on no-load of 12x1.07=12.84V (RMS).

The transformer must be specified. Assuming that the current into the DC load is I$_{DC}$, then for the circuit of **Fig 6.2**(a) the transformer current equals 1.61xI$_{DC}$. For the circuit of **Fig 6.2**(b) the transformer current equals the

DC current. Toroidal transformers create less external magnetic field than their E-I lamination counterparts. They are also less prone to the production of mechanical vibrations.

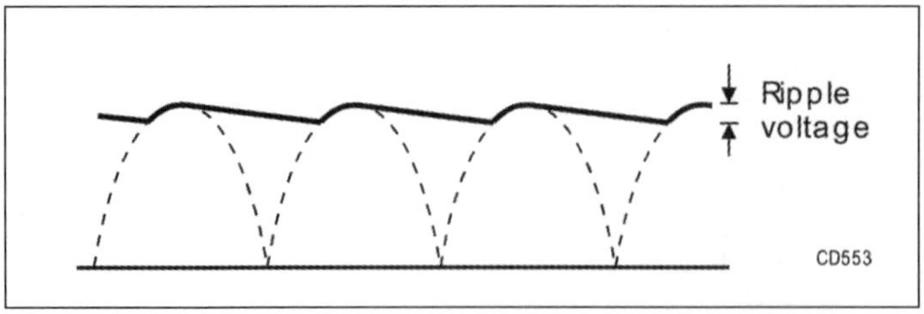

Rectifiers also have to be chosen. Both a voltage and current rating must be determined. In both cases of **Fig 6.3**, the average current per rectifier is approximately 0.5I$_{DC}$. In reality, it is wise to pick a diode with a higher rating than this, especially as the peak charging current of the capacitor can be quite high - so choose a value at least of 1.25I$_{DC}$. The voltage rating of the diodes must also be determined. This should equal at least √2Vs – so choose a value in excess of this. Do not forget that with high current supplies rectifiers usually need a heatsink. In many instances the metal chassis is adequate. The average power dissipation per rectifier will be 0.5I$_{DC}$Vd watts.

The capacitor ratings must now be determined; this requires a capacitance value and a voltage rating. The voltage rating is the easiest and should be greater than the value of the maximum voltage across the capacitor. Electrolytic capacitors are one of the least reliable of electronic components and hence it is wise to choose a voltage value at least 25% greater than the expected maximum voltage, if economically and practically feasible. The capacitance value is a little more complex. The approximate value of capacitance needed to achieve the maximum acceptable ripple is given by:

C = 0.005I$_{DC}$ x 10^6 /Vr µF

Where Vr is the peak-peak ripple voltage, and holds only true only for full wave rectification and a 50Hz supply.

In some applications the basic smoothed supply is adequate, especially if the equipment contains its own regulators.

Stabilised DC Power Supplies

A stabilised supply is one where the output DC voltage is held within close limits

Fig 6.4 Typical output using capacitor smoothing with load

(usually at most tens of millivolts) for quite large variations on the input side. This should take into account working from a low mains voltage (eg 225V); transformer regulation; and smoothing ripple. The regulator is placed after the basic smoothing circuit - see **Fig 6.5**.

Integrated circuit regulators have now become firm favourites. Those that have been around the longest need about a 3V drop across them, whereas the newer breed can manage with less than a 1V drop across them. What is important is that the minimum of any ripple voltage must allow for this voltage drop across the regulator to provide the stable output voltage. If for any reason the ripple is too great, part of it will appear at the regulator output. The familiar Zener diode can be used as a regulator but the price of low power IC regulators in TO92 packages is comparable, and they give better performance.

IC regulators come in two forms - the fixed voltage type and the variable voltage type. They both provide comparable performances. The voltage regulator takes a sample of the output voltage and compares it with an internal reference voltage. It then either increases or decreases the output voltage. Fixed voltage regulators have the sampling accomplished within the IC whereas variable voltage regulators use an external sampling network made up from resistors.

The fixed voltage types come in voltages (both negative and positive) such as 5V, 12V, 15V and 24V. They come also in various current ratings, typical are 100mA, 500mA, 1.5A, 3A and 5A. The only additional components that may be required are decoupling capacitors as specified by the datasheet. The packages range from the plastic TO92

type, 8 pin DIL, plastic TO220, to TO3 types. The variable voltage types will typically provide output voltages between 1.2V and 37V, dependent on type, again either positive or negative voltages. They come in packages as for the fixed voltage types. Apart from possible decoupling capacitors they normally require at least one resistor which is used to determine the output voltage.

There is a multiplicity of regulators available from various manufacturers with a variety of numbers. It is best to look through a supplier's catalogue: remember it is the output voltage and load current that will determine the type. They all generally provide thermal overload protection, short-circuit protection and possibly output transistor safe-operating-area protection.

There are various ways in which a regulator can be used but **Fig 6.6** shows the basic arrangement with capacitors as suggested by SGS and Texas Instruments. Further details are in the relevant data sheets.

To protect a regulator against reverse voltage spikes or where the output voltage decays more slowly than the supply, diodes can be added as in **Fig 6.7**.

→ Good quality components are not cheap, but if used will provide a very reliable power supply which will give good service over the years; conversely, if components of doubtful origin are used, they may not be as reliable.

→ The component ratings used for the power supplies in this section are conservative. It may be possible to reduce ratings of transformers and heatsinks if the power supply is not required to provide the maximum output continuously.

→ For more on power supply design see the *Radio Communication Handbook*, 13th Edn, RSGB 2016, Chapter 23 by Stuart Swain G0FYX

Linear Vs Switched-Mode Power Supplies (SMPS)

Switched-mode power supplies tend to be smaller than a transformer-based power supply for a given power rating, but can be

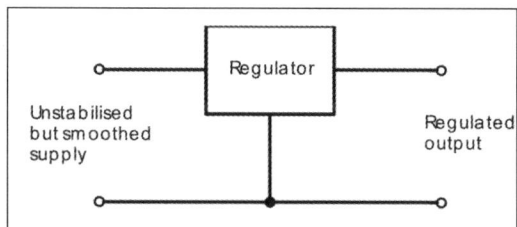

Fig 6.5 Basic regulator arrangement

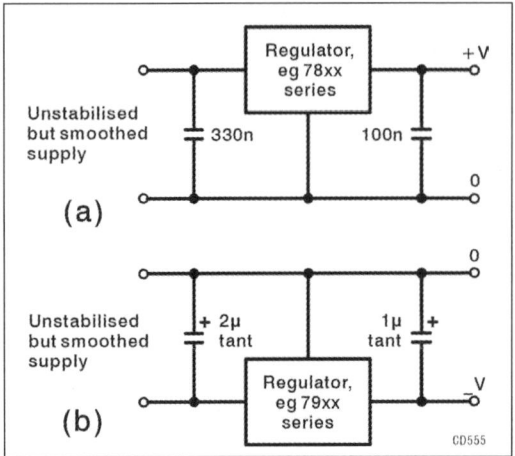

Fig 6.6 Typical regulator arrangement with decoupling - (a) positive output regulator; (b) negative output regulator

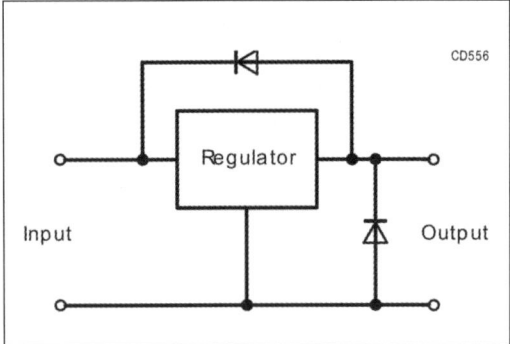

Fig 6.7 Regulator protection

more prone to causing RFI. This is because the input is switched at a frequency typically in the range 50-150KHz, transformed to the voltage required, rectified, and smoothed. The switching waveform is obviously rich in harmonics, but a further problem is that the modulation of the switching waveform causes the nth harmonic to occupy a bandwidth of n times the deviation of the switching frequency. This means that there is potential for interference over a very wide frequency range.

Table 6.1 is a brief comparison for the choice of linear or switched mode power supplies.

→ For an excellent primer on SMPS and

EMC filtering see:

'Design Notes', Andy Talbot, G4JNT, *RadCom*, January & February 2017.

→ For a useful introduction to EMC issues with SMPS see: 'EMC'. David Lauder, G0SNO, *RadCom*, February, 2018.

→ for a short but interesting note specifically on EMC and wireless chargers see:

'EMC', David Lauder, G0SNO, *RadCom*, December, 2017.

→ For issues with noise coming from the voltage regulators, see observation and measurements in: 'Design Notes', Andy Talbot, G4JNT, *RadCom*, March 2018.

→ For a discussion of inter alia, how recent changes in the Wireless Telegraphy Act regulations could impact SMPS see: 'Combating Radio Interference', John Rogers, M0JAV, *RadCom*, June 2016.

Testing Power Supplies

Sometimes power supplies themselves need testing, as when repairing from failure. Generally the following items of test equipment are required:

- Meter for voltage range;
- Meter for current range;
- Oscilloscope;
- Connecting leads;
- Load or resistor-bank.

Note that it is often better to employ a DVM to measure the PSU voltage accurately, but an analogue (rather than a digital) meter or multimeter to measure the current, so that changes and fluctuations in the latter can be more easily observed. The oscilloscope can be used for examining the unregulated and regulated outputs and will show such features as ripple voltage as well as transient response. For switched-mode power supplies, a low-frequency spectrum analyser can also be useful to assess the level of conducted and radiated noise being produced by the switching circuit, which could cause interference. Connecting leads to any supply should be of adequate dimensions to cater for the

Parameter	Linear PSU	SMPS
Size and weight	Transformer for 50 or 60Hz use: tends to be heavy.	Smaller due to higher operating frequency (typically 50kHz–1MHz)
Output voltage	With selected transformer any voltages available. If unregulated, voltage varies significantly with load.	Any voltages available. Voltage varies little with load. Can usually cope with wider variation of input before the output voltage changes.
Efficiency and power dissipation	If regulated, output voltage is regulated by dissipating excess power as heat normally using larger heatsinks. If unregulated, transformer iron and copper losses significant.	Output is regulated using control of duty cycle which draws only the power required by the load. In all SMPS topologies, the transistors are always switched fully on or fully off which keeps heat dissipation to a minimum. Smaller heatsinks.
Complexity	Unregulated may be just a diode and capacitor; regulated will have a voltage regulating IC or discrete circuit and possible protection circuits. Circuits simpler than SMPS.	More complex circuitry - consists of a controller IC, one or several power transistors and diodes as well as a high frequency power transformer, inductors and filter capacitors. Usually mains voltage rectification and smoothing.
Noise and RFI	Normally no high-frequency interference. Some mains hum induction into unshielded cable which can be a problem for low-signal audio. Occasionally mains hum from transformer.	Can be produced due to the high frequency switching of transistors. Increased filtering required to minimize disruptive interference.
Maintenance	Mains voltage likely to be present on primary of transformer and associated switches and fuses. On output side one side likely to be earthed. Mains input side generally checked with a meter. RISK OF SHOCK.	Mains input rectification, smoothing and switching circuits at high voltage – generally no connection earthed. On output side one side of supply likely to be earthed. Difficult to use oscilloscope on mains input side due to no earth present. RISK OF SHOCK.

Supplies with isolation transformers allow a metal chassis to be grounded safely whereas a transformer-less mains-operated supply is potentially dangerous. In both linear and SMPS, the mains, and possibly the output voltages, are hazardous and must be well-isolated.

Table 6.1 Basic comparison between linear and switched mode power supplies

load current. Provision of a simulated load or resistor-bank is dealt with in the next section.

Warning: at all times take care when repairing and testing power supplies, especially when working close to the mains input supply and also on high voltage supplies.

→ For more information see: *Power Supply Handbook*, John Fielding, ZS5JF, reprinted 2009.

Load vs Resistor-bank

Many of the tests to be performed will require a fixed value of load resistance, but the restriction with this is that the current drawn is proportional to the applied voltage (Ohm's law!). In the case of a power supply intended to power an HF transceiver, the test method requires that the load be cycled between two preset levels (receive-only and full-power transmit). This can be done with fixed resistors but it becomes tedious to select and set up resistors when several different load currents are needed.

Where can one get high power resistors? For example a 12V, 10A supply would need a 1.2Ω, 120W resistor for full load testing. This could be made up by four parallel 4R7, 50W resistors in parallel with heavy duty switches, which gives flexibility in providing four values of load current. Looking in various catalogues, a typical wirewound metal clad resistor is ideal but should be mounted on a heat sink, or could also be fan-blown if necessary. An alternative source may be an on-line auction site or a surplus supplier.

Lamps as loads

Take care: the cold filament resistance is often about four to ten times less than the hot value; as a consequence the current drawn when first switched on (the in-rush current) is very high and can cause a power supply current-limiting circuit to latch-up, or in the worst case, fail completely. If a lamp must be used for a load as a temporary measure, insert a low value resistance in series of approximately 25% of the calculated hot value in order to limit the in-rush current.

A Simple Dual-Load Resistor Circuit

This allows a power supply to be cycled between two current levels and provides information about switching transients. The switching device can be an NPN Darlington or a MOSFET. **Fig 6.8** shows an NPN Darlington.

The drive circuit is a simple variable frequency oscillator. The current drawn from a 13.8V supply will be approximately 1A and 10A. The 1.5Ω resistor needs to dissipate high power (about 130W) and the 15Ω resistor needs to be at least a 15W rating. Wire-wound resistors can be used for all the resistors. The 1.5Ω resistor could consist of 10 x 15Ω, 15W wire-wound resistors wired in parallel.

Constant-Current Load

A far better method is to use a simple piece of circuitry which allows the current drawn to be infinitely varied. This has the distinct advantage that once the current has been set to a value it remains constant, even if the supply output voltage changes for any reason. This is known as a 'constant-current load' or 'current sink' and can be made with a few power transistors and an op-amp.

The power dissipation can be anything desired, assuming correct choice of the heatsink and power transistors. A slight modification to the basic circuit allows injection of an audio signal to 'modulate' the load current. This audio signal can be anything from a few Hertz to many kilohertz and can be either sinusoidal, triangular or a rectangular wave. With such an item of test equipment one can fully explore the transient response of low voltage/high current power supplies.

→ See Chapter 8.15 for construction of a home-brew current sink.

Thermal tests

Overheating will seriously degrade the reliability of a semiconductor and generally lead to its demise. So it is wise to keep the temperature of a semiconductor as cool as possible. A very rough rule of thumb is never to let the temperature of the case of a transistor (when mounted on a heatsink) exceed what can be tolerated on the skin on the outside of the first finger between the two knuckles (about 45°C) - remember the chip inside the semiconductor will be much hotter.

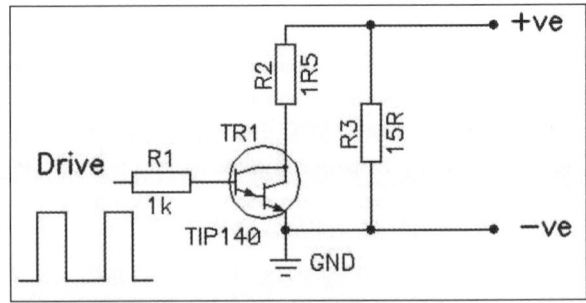

Fig 6.8 A simple dual load resistor circuit for transient testing

Monitoring the case temperature of a semiconductor is not difficult – the one warning is to beware that the case may be above ground potential. Use a DVM with a temperature probe, or a max/min thermometer with an external probe, and apply heatsink compound for good thermal connection to the probe wherever possible. A safer option could be to use an infrared probe if one is to hand.

Discharge probe

When accessing the inside of a power supply disconnected for maintenance, it is good practice to use a probe to discharge large-value capacitors. These capacitors may pose a hazard by virtue of their residual charge, and the consequent high voltage across them, or the high current that they can deliver if shorted. This type of probe may include a series resistance to restrict the current, so that electrolytic capacitors are not discharged too quickly.

Such a probe can also be used in principle to keep a capacitor discharged while the equipment is being maintained, and to reduce the risk to the operator if the equipment is accidentally switched on. But beware! 'strapping across' the circuit requires extremely careful consideration of the probe characteristics (especially its resistance and dissipation) and of the consequences if the equipment were to be accidentally powered up.

→ See Chapter 8.1 'Probes' for how to make one.

Transformer Testing

Occasionally it may be necessary to investigate a transformer, perhaps to determine its turns-ratio and/or the phasing relationship of its windings.

Set an oscillator to the test frequency and connect it to the primary of the transformer under test, applying 1V as indicated on an AC voltmeter or oscilloscope (**Fig 6.9**). Measure CD and read the voltage. The value read

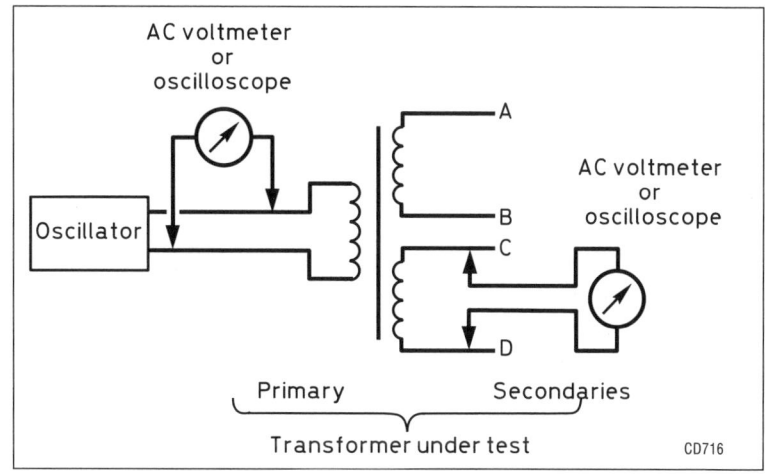

Fig 6.9 Transformer testing

is the turns-ratio. It can then be repeated for winding AB.

To check the phase relationships of the secondary windings, link BC and measure the voltage across AD. If this equals the sum of the voltages in the turns ratio test, then A and C can be considered as the start of the windings. If not, connect C to A and measure across BD, this should then equal the sum of the voltages taken in the turns ratio test, then B and C can be considered as the start of the windings. If the phasing of more windings is required continue in a similar manner. This test cannot determine the voltages or currents at which the transformer is designed to operate, nor can it determine the number of turns of a winding, however, for a mains transformer, the voltage rating of the primary-winding is often known.

7 System Measurements: Specific Methods

7.1 Transmitters

THE UK Amateur Radio Licence (at section 9(2)) requires the licensee to operate within the prescribed frequency bands, and places a restriction on the maximum level of power that can be transmitted in those bands by the station depending on the class of licensee and the band in use (Schedule 1, Tables A-C). Furthermore, the licence (at section 7) requires the licensee to ensure that the station does not cause undue interference to others, by factors such as unwanted emissions and excessive bandwidth. This is summarised and reinforced by section 7(5), which requires the licensee to, 'conduct tests from time to time to ensure that the requirements set out in this Clause 7 are met'.

Clearly, the licensee has a duty to check that his/her emissions are within the required limits, and this in turn implies that they have the expertise and equipment to do so. It is a legal requirement (not an option) to comply with these requirements – they are there for a reason, and it is not clever to ignore them. Non-UK Amateurs will doubtless have similar responsibilities. This section is therefore devoted to measurements associated with checking for such compliance.

Transmitted Power

The maximum power level permitted in each band is defined by Schedule 1 in terms of PEP. In a couple of cases (at VLF and UHF) the peak ERP is specified. The notes to Schedule 1 tell us that:

• Peak envelope power (PEP) is the average power supplied to the antenna by a transmitter during one radio frequency cycle at the crest of the modulation envelope taken under normal operating conditions.

• Effective radiated power (ERP) (in a given direction) is the product of the power supplied to the antenna and its gain relative to a half-wave dipole in a given direction.

→ For more on power meter measurements see Chapter 5.5.

CW

In a carrier wave (CW) situation, or with a frequency modulated signal, the output is of constant amplitude, and so it is relatively easy to measure the output power. Simply key the transmitter and determine the RMS voltage (Vrms) of the resulting carrier across a dummy load (R). The power is given by:

$$P = Vrms^2/R \quad \text{watts}$$

AM

If the signal is amplitude modulated (double-sideband with carrier), the overall output power increases. The power is divided between the sidebands and the carrier component. With 100% modulation (modulation index=1), the output power increases to 1.5 times the unmodulated condition where the power contained in each of the two sidebands is one quarter that in the carrier. For this form of modulation it is convenient to measure the carrier power (AM with no modulation, or in other words CW carrier) as described above. This value can be multiplied by 1.5 to give the maximum output power available.

If an exact value of output power at a particular level of modulation is required, it is necessary to determine the modulation index (m). The output power is then given by:

$$P_{out} = (1+\tfrac{1}{2}m^2)V^2/R \quad \text{watts}$$

where V is the RMS value of the unmodulated carrier and R the load.

SSB

With single sideband modulation, there should be no power output until modulation is applied, but when it is applied, the mod-

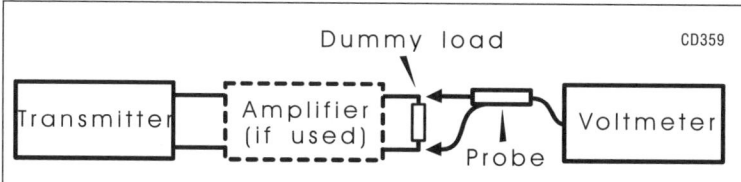

Fig 7.1 RF power measurement using a probe and voltmeter

ulation envelope is non-sinusoidal in appearance. The normal method of measuring output power is by observation of this envelope and determination of the peak envelope power (PEP) - the parameter defined by the licensing authority as noted above.

Power Measurements Using RF Voltmeters or Probes

An RF voltmeter, or a probe used in conjunction with a voltmeter, can be used provided it covers the power and frequency to be measured, and should be calibrated against a power meter of known accuracy. If a peak-reading voltmeter is being used, do not forget to convert peak to RMS values by dividing by $\sqrt{2}$ before using the basic formula given in the previous section. This method of measuring power should be used only for constant carrier amplitude signals, such as unmodulated AM, CW, and FM. **Fig 7.1** shows the basic arrangement.

Power Measurements Using the Oscilloscope

The oscilloscope may be used up to about 30MHz to monitor modulated waveforms and measure output power, but at higher frequencies the oscilloscope becomes an expensive item, and may provide unwanted loading effects on the equipment being monitored.

The most straightforward approach, as with the RF voltmeter, is to measure carrier power with the key down in CW mode, or the constant amplitude of a frequency modulated signal. Connect an os-

cilloscope instead of a voltmeter (**Fig 7.1**) bearing in mind any frequency or voltage limitations of the oscilloscope and probe, and measure the peak-to-peak amplitude V_{pp} across the known dummy load, R (**Fig 7.2**).

The average power is then calculated from:

$$Pavg = V_{pp}^2/8R \text{ watts}$$

For PEP measurements, make the same physical connections across the dummy load with the oscilloscope, but drive the SSB transmitter from a two-tone oscillator (**Fig 7.3**). The output of the oscillator should be fed into the microphone socket with an amplitude equivalent to that from the microphone. Set the timebase on the oscilloscope to be in the audio range, and a waveform similar to that shown on **Fig 7.4** should be obtained. Measure the peak-to peak-voltage V_{pp} as shown, and calculate the power using the same formula as above.

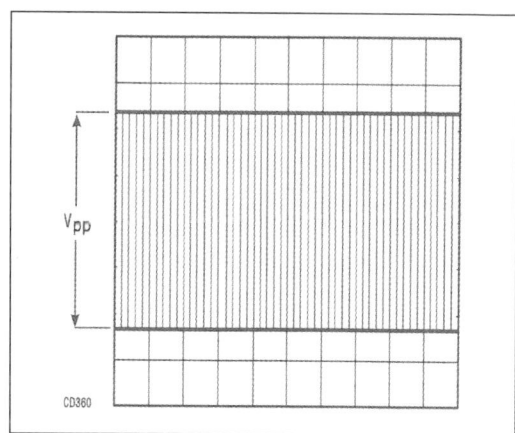

Fig 7.2 Oscilloscope display of carrier only

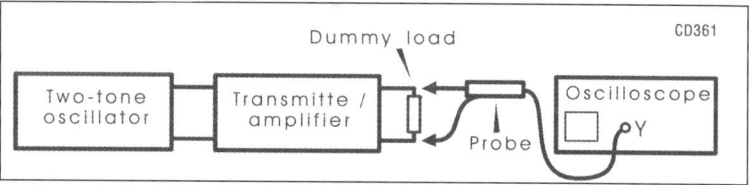

Fig 7.3 RF measurement for SSB work

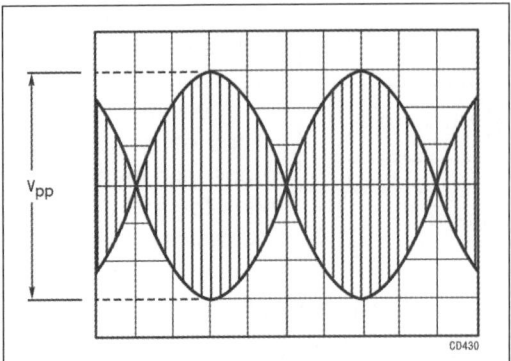

CD430

Fig 7.4 Two-tone test display

If the same oscilloscope is continually used to monitor output power on SSB, marks could be made on the graticule or display, of the positions corresponding to various power levels. The peak of the speech modulated waveform should then never exceed the maximum permitted level (**Fig 7.5**).

→ Safety note: At 100W (CW) the peak-to-peak voltage across a 50Ω dummy load is 200V; at 400W PEP the maximum peak-to-peak voltage will be about 400V. Be careful!

→ Warning: The capacitance of an oscilloscope can start to have an appreciable effect at 30MHz. A reactance of 25pF is 212Ω at 30MHz, and obviously affects the readings. It may be better to use a divide-by-ten probe to decrease the parallel capacitive loading to about 12pF. This still represents a capacitive reactance of 442Ω at 30MHz and the voltage read from the screen will be lower than the

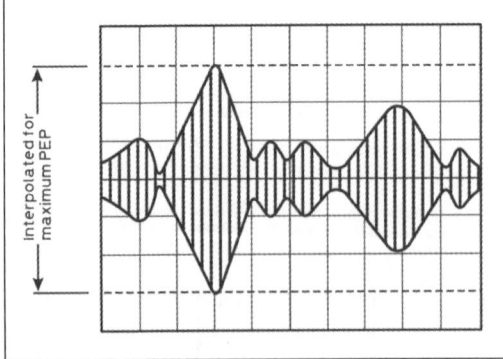

Fig 7.5 Speech waveform and interpolated maximum PEP level

real value - this loading is all in parallel with the dummy load.

→ For a really interesting article on ways of measuring RF power, see:

'Homebrew', Eamon Skelton, EI9GQ, *RadCom*, February & March 2017.

Modulation Checks (See also Chapter 3.1)

A 'sniff' of the transmitted signal should be obtained using of an RF sampler or coupler. A direct connection method is shown in **Fig 7.6**, where the potential divider formed by a 9k1 and 1k resistive network provides an output voltage which is one tenth of that across the load. This assumes that the oscilloscope presents a high (and therefore effectively negligible) input impedance to the network. Providing carbon resistors are used, this method is usable up to the limits of the oscilloscope. The resistors are 5W rated (corresponding to an RF output of 400W).

It does not matter if the exact sample ratio is not known as this measurement is over a relatively narrow bandwidth and the measurements are relative to one another. **Fig 7.7**

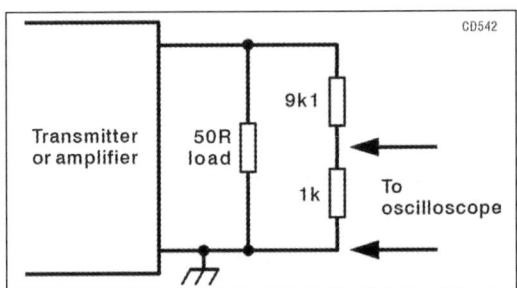

Fig 7.6 Direct connection

shows another method whereby an insulated coil connected to the oscilloscope can be positioned somewhere in the vicinity of amplifier output components to couple some output..

Modulation Depth (A3E) - Method 1:

To find the AM modulation depth, a modulated RF is applied to the Y-input of an oscilloscope, the timebase being adjusted to give a waveform similar to that shown in **Fig 7.8** for single frequency sinewave modulation.

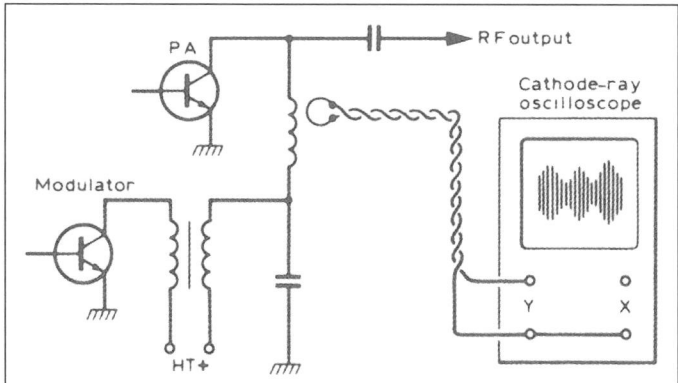

Fig 7.7 Arrangement with movable coupling loop

As it is easiest to measure peak-to-peak values on an oscilloscope, the percentage modulation is given by:

100(A-B)/(A+B)

Fig 7.9 shows various conditions of modulation.

Modulation Depth (A3E) - Method 2:

This method does not require a timebase. Instead, the modulating audio frequency is

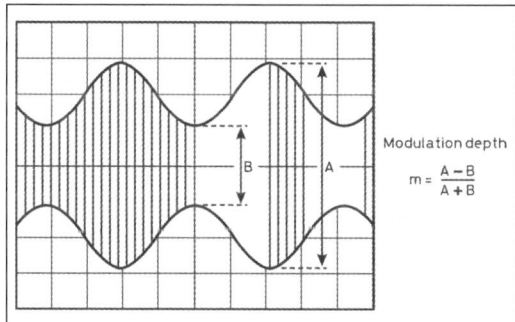

Fig 7.8 Modulation depth measurement

fed to the X-input and a sample of the modulated output is fed to the Y-input - as shown in **Fig 7.10**.

Fig 7.11 shows some typical waveforms that can be expected. The vertical line at (a) shows the unmodulated carrier amplitude; in (b) the carrier is modulated 50% while in (c) it is modulated 100%. Where the sloping edges of the pattern are flattened as in (d), the carrier is over-modulated. The modulation depth is given by the above formula.

If more convenient, the circuit shown in **Fig 7.12** can be used. Here LC tunes the RF sig-

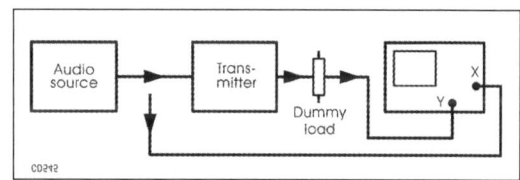

Fig 7.10 Amplitude modulation - method 2

nal to produce the Y-deflection, and detected transmitter audio provides the X-deflection.

Amplifier Linearity - General

When a post-transmitter amplifier is used with amplitude modulation, linearity is of the utmost importance, otherwise distortion of the modulation may occur. This distortion in turn can cause unnecessary interference to other users. A very simple check is to take a sample of the output of the amplifier/transmitter and display it on an oscilloscope with internal timebase. **Fig 7.13** shows possible oscilloscope waveforms. Waveforms at (a) and (c) represent no apparent distortion whilst those at (b) and (d) show the amplifier being overdriven.

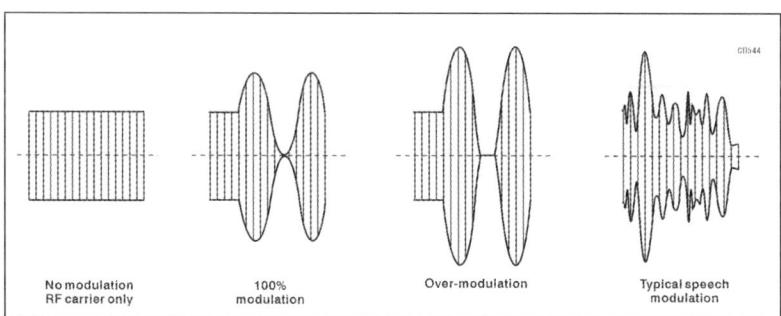

Fig 7.9 Various conditions of amplitude modulation

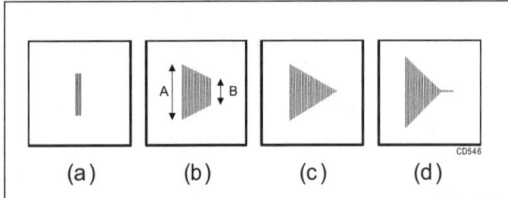

(a)   (b)   (c)   (d)

Fig 7.11 Typical oscilloscope patterns obtained using method 2

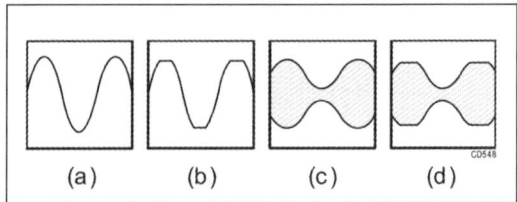

(a)   (b)   (c)   (d)

Fig 7.13 Possible oscilloscope waveforms

Amplifier Linearity - Single Sideband (J3E)

The linearity of an SSB amplifier may be checked by the 45° method. This is shown in **Fig 7.14**, which also includes a suitable sampling circuit (b). A sample of a detected signal is taken from before and after the amplifier and applied to the X and Y inputs respectively. The resistors R1 and R2 of the sampling circuit should be adjusted so that the voltage output from each detector is similar. Referring to **Fig 7.14** (c) - the ideal display should be a single line at 45° (1); possibly incorrect bias is shown at (2), (3) shows amplifier overloaded, (4) shows insufficient standing current in the amplifier whilst (5) is the pattern obtained from speech input to a correctly adjusted amplifier.

Using the arrangement as shown in **Fig 7.15** (a), the modulation pattern can be monitored. Typical two-tone test outputs are also shown on the same diagram (b). The pattern at (1) shows an amplifier correctly adjusted, in (2) the peaks are flattened due to insufficient loading or over-driving, and at (3) distortion at crossover points due to incorrect bias. If, on removing the audio modulation, there is still some trace of a carrier shown on the screen, then this indicates that the carrier null may need adjustment.

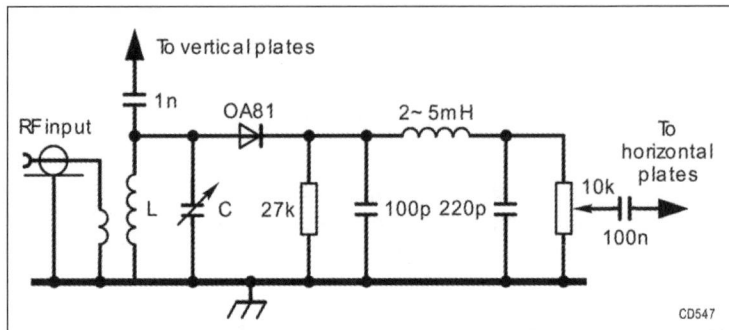

Fig 7.12 Alternative method of obtaining modulating signal

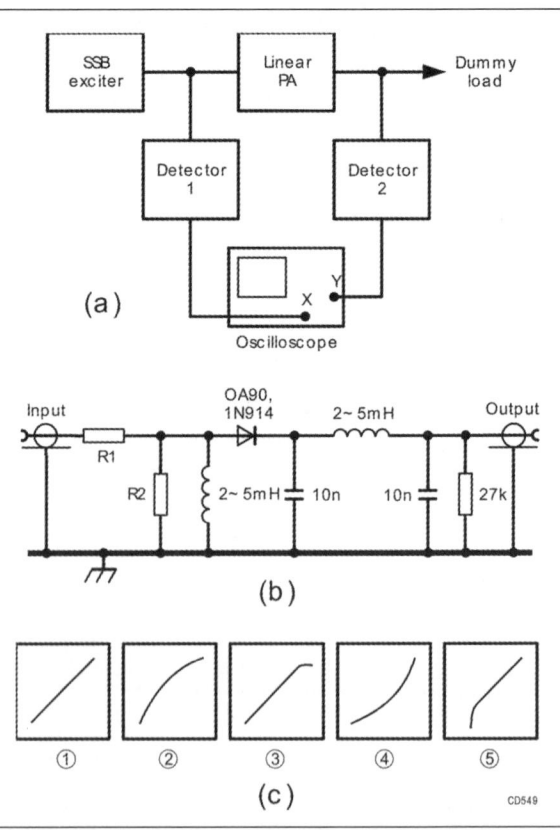

Fig 7.14 45° method for checking amplifier linearity

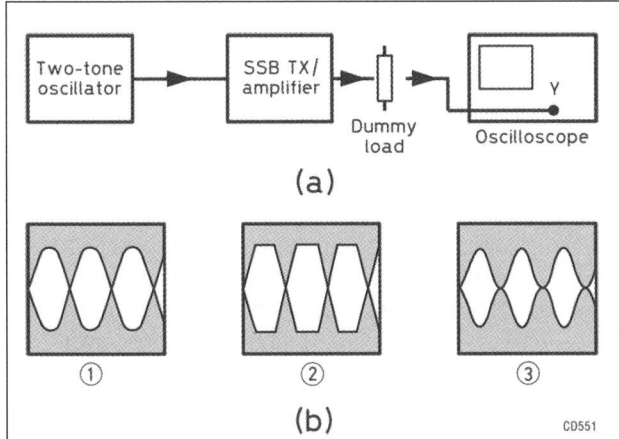

(a)

(b)

CD551

Fig 7.15 Two-tone amplifier / transmitter testing

Frequency & Spectral Purity (See also Chapter 3.2)

The measurement of frequency is of prime importance to the Radio Amateur in order to ensure that transmission occurs only within the authorised bands and hence complies with the licence conditions. Although synthesisers are commonplace in radio equipment, there are still transmitters using variable frequency LC oscillators.

With synthesisers, the digital display is usually driven by the binary digits that set up the division ratios and *not* the actual frequency produced. It is therefore imperative that highly accurate frequency measurements can be made - to a higher accuracy than any other measurement that the Radio Amateur has to make.

Spectral purity (aka 'a clean signal') is important so as not to cause undue interference to other users of the spectrum. This involves not just the exciter, but any following amplifiers used.

CW Measurements

The test arrangement for carrying out CW transmitter measurements is shown in **Fig 7.16**. The measurement of power, harmon-

ic, and spurious outputs is largely self-explanatory. It is important to keep input attenuation levels on the spectrum analyser as high as possible to avoid the possibility of over-driving it, which would give inaccurate harmonic measurements due to spurious generated within the analyser itself. A critical check therefore, is to insert an additional 10dB attenuation and make sure that the relative levels do not change, just a change in the noise floor.

CW keying is checked at 40 words per minute, which gives a 31ms dot length on the oscilloscope. Rise time, fall time, delays or distortions and any first-character differences are all noted, at both full break-in and semi-break-in.

SSB Measurements

The test arrangement for carrying out SSB transmitter measurements is shown in **Fig 7.17**. First, the transmitter is driven to full rat-

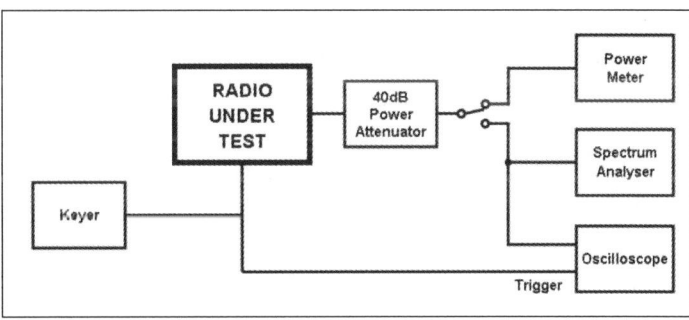

Fig 7.16 CW transmitter measurements

ed output power using a single audio tone and the amplitude of the waveform displayed on the oscilloscope noted. Then the transmitter is driven by two equal-level audio tones (700Hz and 1700Hz) to the same peak amplitude on the oscilloscope. The PEP level is then the same as the power output on a single tone. However, if an accurate and reliable PEP meter is available, the oscilloscope transfer method is not needed and the power level can be read directly from the meter scale.

The level of the 3rd and 5th order intermodulation products is measured using a spectrum analyser. It is common to quote levels with respect to PEP as this is universally adopted for all amateur radio products, reviews and specifications. With the transmitter driven from a single audio tone, the levels of carrier and unwanted sidebands are measured using the spectrum analyser, and an estimate made of audio distortion.

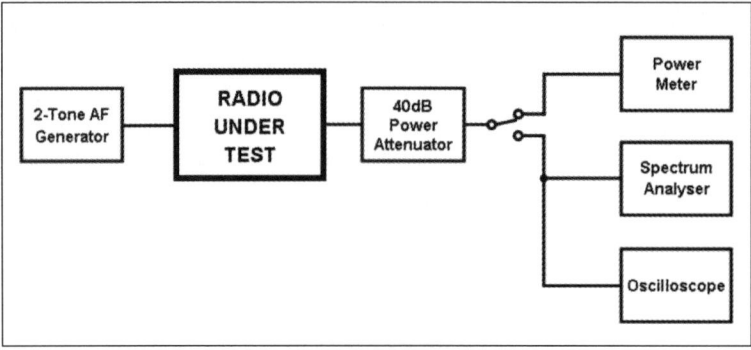

Fig 7.17 Two-tone transmitter measurement

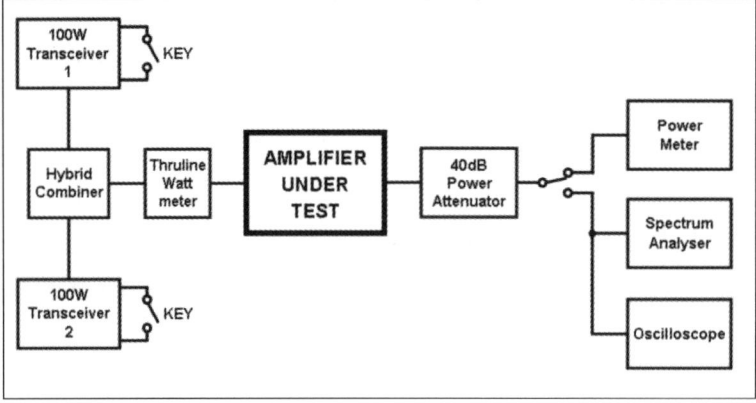

Fig 7.18 Two-tone linear amplifier transmitter measurements

Linear amplifiers

Fig 7.18 shows a test arrangement for measuring the two-tone intermodulation distortion performance of an SSB linear amplifier. Here, two 100W HF transceivers operating at slightly different frequencies are coupled together using a high-power combiner and fed into the linear amplifier under test. The residual level of intermodulation products should be around -50dB to -60dB, so that the distortion in the drive source is significantly better than that of the amplifier to be measured.

→ Care! The transceivers are used in CW mode, and both must be switched to transmit before either key is pressed. Also both keys must be released before either transceiver is switched back to receive. This is most important otherwise leakage through the combiner from one transmitter could damage the receiver in the other transceiver.

→ Homebrew designs for high-power combiners using ferrite-cored transformers are described in books on RF design.

Transmit-Receive Switching Speed

A reasonably fast and a clean switch-over between receive and transmit and back to receive is needed on some data modes. The test arrangement is shown in **Fig 7.19**. An audio generator at 1kHz drives the transmitter in AFSK mode via the AFSK or microphone input. The receive signal generator is set to give a reasonable on-tune signal level and coupled into the antenna connector via 60dB of attenuation. This ensures that the amount of transmit signal entering the signal generator is insufficient to cause any damage. The data PTT line on the radio is keyed from a pulse generator and the resulting RF output, receiver audio and keying waveform observed on a multichannel oscilloscope. The times for the receiver to be muted, the transmitter to reach full power, the transmitter to be muted and the receiver to regain full sensitivity are all read from the oscilloscope display, and any

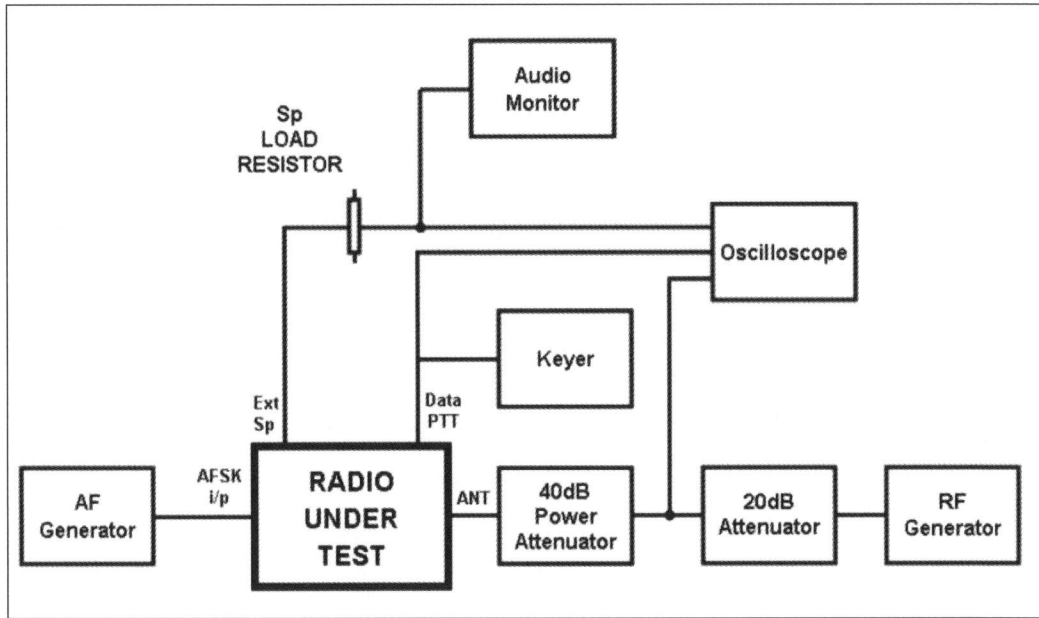

Fig 7.19 T/R switching speed measurement

anomalous behaviour can be observed. Receiver and transmitter mute times are usually very fast; transmit and receive enable times less than 20ms are quite fast and fully acceptable for all modes.

7.2 Receivers

Many of these measurements require quality test equipment which only a few will either own or have access to. Some of this equipment is obtainable at radio rallies, surplus stores or on Internet auction sites.

→ A good source of reference, and with further details, is Peter Hart's book, specifically the first chapter: *25 Years of Hart Reviews*, Peter Hart, G3SJX, RSGB, 2005.

There are broadly two sets of parameters which define the effectiveness of a receiver: how well it receives wanted signals, and how well it rejects unwanted signals. Key receiver

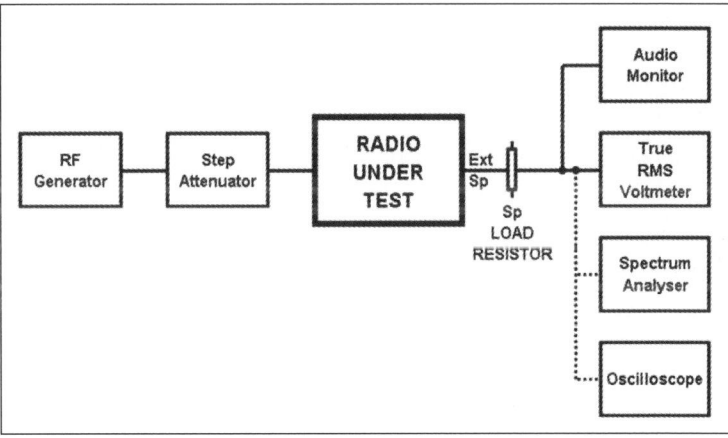

Fig 7.20 Sensitivity related receiver measurements

measurements might therefore include:

- Sensitivity and SINAD;
- Noise figure;
- AGC response;
- Intermodulation and blocking;
- Receiver reciprocal mixing.

These parameters are examined in turn below

Sensitivity

There's a choice of methods for expressing this.

Signal-to-Noise

Fig 7.20 shows the test arrangement for making measurements of sensitivity, spurious response rejection and selectivity. A suitable resistive load is connected to the external loudspeaker socket of the radio and the audio output monitored using a true RMS voltmeter. It is important that this indicator can show the true RMS level of the noise output as well as a mixed noise and sinewave output. A dB scale is by far the most convenient. The audio output level should be low enough to avoid audio distortion at all times. The audio monitor shown in **Fig 7.20** is just a simple amplifier and speaker so that the receiver audio can be heard.

As an alternative to the voltmeter, an audio spectrum analyser or FFT analyser can be used with software to compute signal-to-noise ratios directly. This can have a number of advantages with improved accuracy and ease of separating signals and noise.

Sensitivity measurements on SSB or CW are made with the RF generator set to give a 1kHz audio beat note. With the RF generator switched off, the audio output level on noise alone is set to a convenient value. The generator is then turned on and the RF level set to give a 10dB increase in audio output. This level gives the sensitivity for 10dB (signal+noise) to noise ratio (s+n):n. The receiver noise floor is 9.5dB lower than this figure. If the generator had been set to give a 3dB increase in audio output, the level would be equal to the noise floor of the receiver. Measurements of (s+n):n for AM and FM signals are made, not by turning the RF generator on and off, but by switching the 1kHz modulation on and off: 30% modulation depth is usually used for AM and 3kHz deviation on FM.

SINAD

This is a measurement that can be used with any radio communications equipment and examines the degradation of the signal by unwanted or extraneous signals including noise and distortion. It is a common performance measurement in many applications including many two-way radio FM communications systems especially at VHF and above.

SINAD is defined as the ratio of the total signal power level (Signal+Noise+Distortion) to the unwanted signal power (Noise+Distortion). That is:

$$\text{SINAD} = 10 \log_{10} \{(S+N+D)/(N+D)\} \text{ dB}$$

It is typically used for measuring and specifying the sensitivity of a radio receiver and is normally expressed in decibels (dB). The higher the figure for SINAD, the better the quality of the audio signal. Note that SINAD is a *power* ratio, and not a *voltage* ratio in this calculation.

Fig 7.21 shows a typical arrangement for measuring SINAD. An RF signal modulated with a 1kHz audio tone is fed into the radio receiver. Two audio measurements are then made, one with the 1kHz modulation for (S+N+D), and one with it notched out for (N+D). Once the figures are obtained, a value for the SINAD of the radio receiver can be calculated.

The notch filter characteristics for SINAD measurements are of importance and ETSI

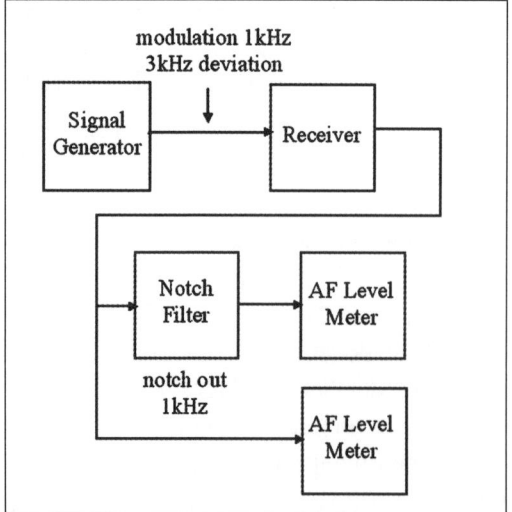

Fig 7.21 Arrangement for SINAD measurement

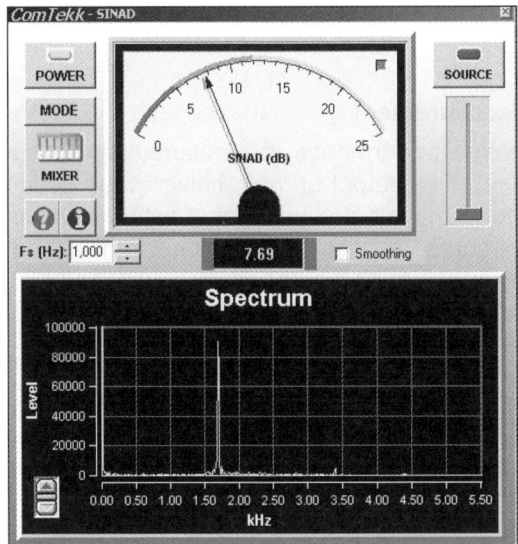

Fig 7.22 Software SINAD meter display

the spectral content of the received audio signal. The receiver under test is fed with a calibrated RF signal from a signal generator. This signal is modulated to about 60% of maximum with a 1kHz sine wave tone (usually FM). The receiver audio output is coupled to the computer's sound card where an A/D converter samples the signal's instantaneous voltage. With time and magnitude data, the software first processes each batch of samples by applying a bandpass filter or window, thereby ignoring any frequencies outside the typical communications range of 300 to 3400Hz. Another filter separates the 1kHz signal from the rest which is considered as noise. A screen shot is shown in **Fig 7.22**.The program typically measures SINAD, Signal-to-Noise Ratio (SNR), Total Harmonic Distortion (THD) and THD+N. See *http://comtekk.us/sinad.htm*

(the European Telecommunications Standards Institute) defines such a notch filter in ETR 027. Briefly, this specifies that for a standard modulating frequency of 1kHz, the 1kHz signal should be attenuated by at least 40dB, at 2kHz the attenuation should not exceed 0.6dB, and the filter characteristic shall be flat within 0.6dB over the ranges 20Hz to 500Hz and 2kHz to 4kHz. Without modulation, the filter shall not cause more than 1dB attenuation of the total noise power of the audio frequency output of the receiver under test.

Whilst measurements of SINAD can be made by configuring a number of individual items of test equipment, there are purpose-built commercial SINAD meters (and kits to build one) which can make testing easier. Better still perhaps, these days a suitable program can be downloaded onto a PC, to convert it into a SINAD meter; the receive audio fed into it to the PC, and the program computes SINAD (and usually other parameters as well) by means of digital signal processing (DSP).

For example, a Windows 10 compatible program from ComTekk uses DSP technology to measure some key receiver performance parameters, including SINAD, by analysing

Sensitivity performance of a radio receiver can be assessed by determining the RF antenna input level needed to achieve a SINAD figure of 12dB (a typical figure that is used commercially). This equates to a distortion factor of 25% with a modulating tone of 1 kHz. Note that ETSI typically specify a deviation level of 12.5% of the channel spacing; with AM the modulation depth needs to be specified.

A typical specification for a narrow band FM receiver might state that a receiver has a sensitivity of 0.25µV for a 12 dB SINAD. The lower the RF input voltage needed to achieve the given level of SINAD the better the receiver performance. A 12dB SINAD figure is considered the maximum acceptable level of noise that will not swamp intelligible speech.

Noise Figure

As an alternative to measuring (S+N):N or SINAD, the noise figure of a receiver may be measured using a suitable calibrated noise source. This technique gives more accurate results, particularly with low-noise VHF,UHF, and microwave receivers, and has the advantage that noise figure is independent of receiver bandwidth.

Theory

In any amplifier, noise is added to the signal so that the signal-noise ratio at the output of the amplifier is worse than at the input, even though the signal has been amplified. This is especially important in receiver RF amplifiers which deal with low level signals.

The ratio of the signal-to-noise at the input, to the signal-to-noise at the output, defines the Noise Factor. As this ratio can have a wide range of values, it is convenient to express it in decibels (dB) and call it the Noise Figure (NF). Hence:

$NF = 10\log_{10}\{(s/n\ in)/(s/n\ out)\}$ dB

In a receiver, there are several cascaded stages each of which will contribute noise, but the effect of the noise contribution of each successive stage is reduced by the power gain of the preceding stage. Thus if NF_1, NF_2, and NF_3 are the respective noise figures of each successive stage and G_1, G_2 and G_3 are the stage gains, the overall receiver noise figure NF_R will be given by:

$NF_R = NF_1 + \{(NF_2-1)/G_1\} + \{(NF_3-1)/(G_1G_2)\}$

In most cases only the first and second terms are significant.

Receiver noise figure (NF_R), noise floor, and bandwidth may be related for an SSB receiver as follows:

Noise floor (dBm) = $NF_R + 10\log_{10}$ (bandwidth in Hz) -174

Measurement

Noise performance is measured by noting the noise output of the receiver when its input terminals are terminated with the value of source resistance for which it is designed and then adding a known amount of noise at the input such that the value of output noise is doubled. It is then obvious that the added noise is equal to the noise generated by the receiver, although two assumptions are made for this to be true: firstly, that all of the known output from the noise source is, in

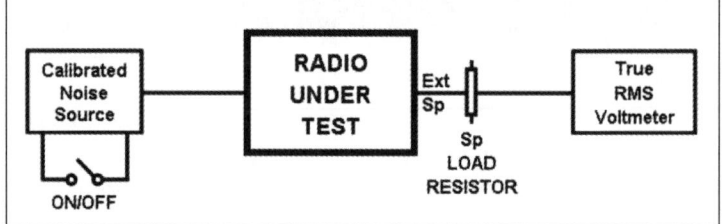

Fig 7.23 Noise figure measurements

fact, coupled into the receiver, and secondly that the receiver output doubles when the effective input is doubled (ie the receiver is linear over this range of inputs).

The first point will be met provided that none of the noise is shunted, that transit time effects are negligible, and that the output of the noise source is coupled into the receiver by a

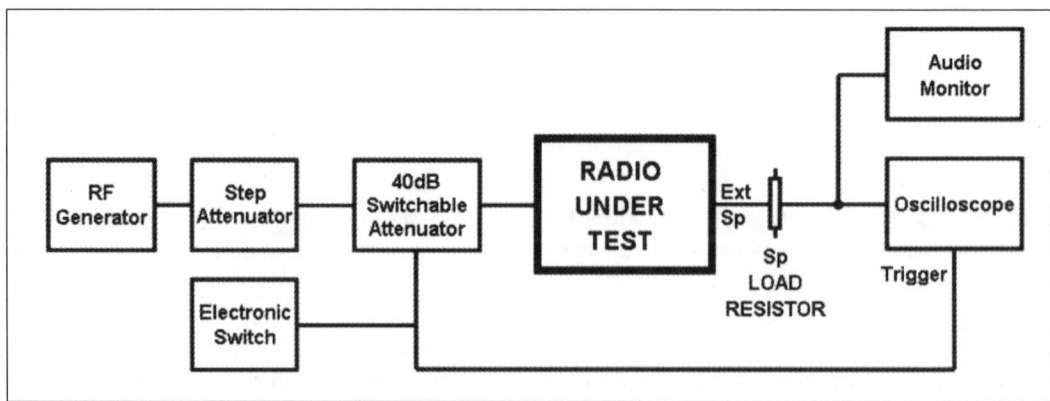

Fig 7.24 AGC measurements

very short length of low-loss cable of the correct characteristic impedance. The linearity of the receiver can be established by providing two identical noise sources and shunting the output of the receiver by a 3dB attenuator when the second source is switched on, but for amateur use this is scarcely worthwhile.

Fig 7.23 shows a typical arrangement of equipment. The measurement requires a noise source whose ENR is known at the frequency of interest, and whose impedance matches the receiver input impedance. The noise source impedance should not change between its on and off states as this will affect the measurement. An audio power meter may be used, or a voltmeter, provided that the impedance across which the audio output power is measured is known so that power can be calculated if required.

If the noise source is variable, it is set to give a 3dB increase in noise level and the receiver noise figure is simply read from the calibrated scale on the instrument.

With a fixed noise source, the noise generator is turned off and the receiver audio gain adjusted for a convenient noise reading on an audio power meter (N_1). Next, the noise generator is turned on, and the new output (N_2) observed. The resulting power ratio, N_2/N_1 is referred to as the Y-factor, and this noise figure measurement is commonly called the Y-factor method.

The noise figure can be calculated from:

$$NF = ENR - 10\log_{10}(Y-1) \text{ dB}$$

Where ENR is the Excess Noise Ratio of the noise generator in dB, and is given by:

$$ENR = 10\log_{10}\{(P_2/P_1) -1\} \text{ dB}$$

Here P_2 is the noise power of the generator and P_1 is the noise power from a resistor at 290°K. Calibrated noise sources usually have their ENR marked in the side of the noise head.

→ See 'The Measurement of Noise', Dave Roberts, G8KBB, *RadCom* January 2007

→ Noise figure measurements, Chapter 12, *The ARRL Handbook for the Radio Amateur (2010)*

→ Appendix B, Section 1 for more on the theory of noise measurement

AGC Response

Fig 7.24 shows the test configuration for making AGC measurements. The drive source from the generator includes a home-constructed 40dB attenuator which can be switched quickly and cleanly in or out of circuit by an electronic trigger signal. This signal is also used to trigger an oscilloscope scan to view the audio output from the receiver. The attack and decay characteristics for a 40dB change in level over a range of signal inputs can be observed together with the threshold level at which the AGC starts to operate.

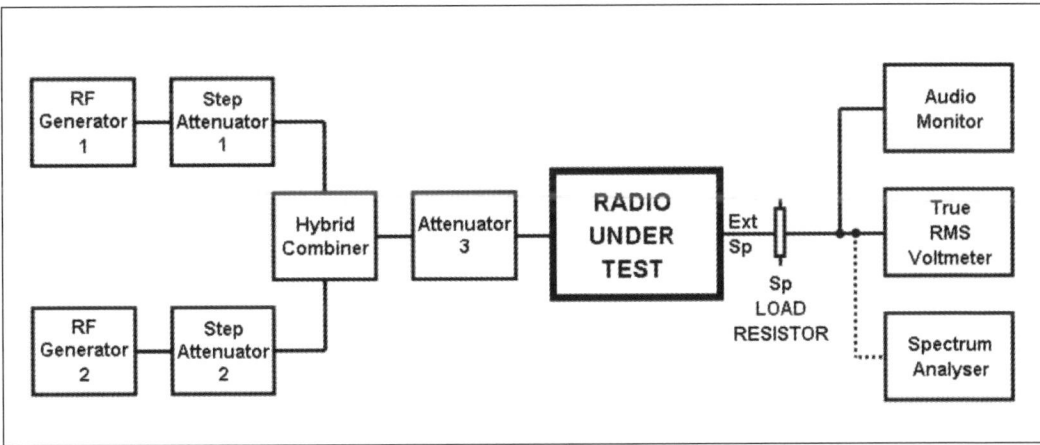

Fig 7.25 Intermodulation and blocking receiver measurements

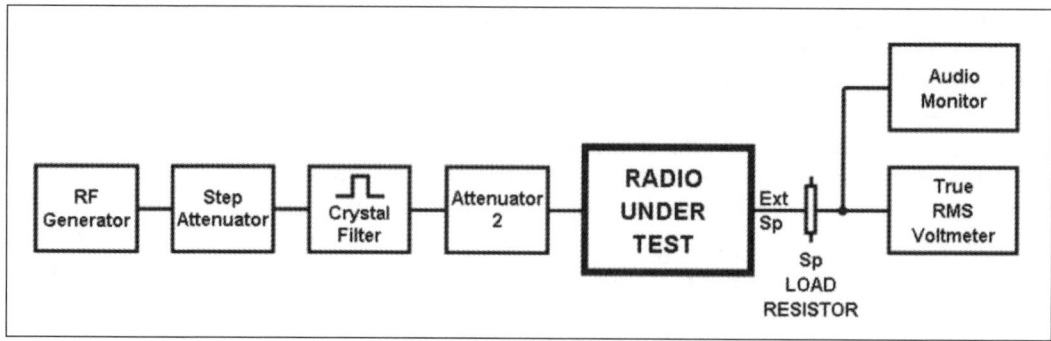

Fig 7.26 Reciprocal mixing measurements

Intermodulation and Blocking

The test arrangement for making measurements of intermodulation and blocking is shown in **Fig 7.25** and requires two signal generators.

In making intermodulation measurements, it is most important to ensure that intermodulation products are not introduced by one generator coupling into the other. A hybrid coupler and not just a resistive combiner is necessary as this will give some 30dB extra isolation. This extra isolation will only be achieved if the coupler is properly terminated at the output port and this requires that attenuator 3 in **Fig 7.25** is not reduced too low in value - a 10dB attenuator is typically used.

Even with a suitable coupler, it is quite difficult to keep test equipment intermodulation within limits and get accurate results when receiver dynamic ranges exceed 100dB. Some generators are better than others and a careful balance between generator output is often necessary to achieve best results.

To measure third-order intermodulation, the levels of the two generators are increased equally by using step attenuators until the amplitude of the intermodulation product generated in the receiver gives a measured s+n:n ratio of 10dB. Note: if the AGC is operative at this level, a spectrum analyser or notch filter needs to be used, as for sensitivity measurements under this condition.

The difference between the amplitudes of either input signal as measured at the antenna input, and the on-tune level or sensitivity figure for similar s+n:n, is termed the intermodulation ratio. However, it is more convenient to quote intermodulation performance in terms of 3rd-order Intercept Point (IP), as this is independent of measurement signal-to-noise ratio and bandwidth. That is:

IP (dBm) = (3S-I)/2

where S is the amplitude in dBm of each input signal and I is the amplitude in dBm of the intermodulation product generated when related to the receiver input.

Having measured the third order Intercept (IP_3) and the noise floor (N_{Floor}) of the receiver from the sensitivity measurement, the two-tone spurious free dynamic range (or intermodulation-limited dynamic range) is calculated from the expression:

Dynamic Range (dB) = $0.667(IP_3-N_{Floor})$

where IP_3 and N_{Floor} are expressed in dBm.

For every 1dB increase in generator levels, the third-order intermodulation products in the receiver increase by 3dB. If this does not happen, intermodulation may be occurring simultaneously in more than one stage in the receiver, front-end AGC may be operative, or intermodulation products may be generated within the measuring set-up. The above expressions for IP_3 and dynamic range assume that the 3:1 ratio holds and cannot be accurately applied if this is not the case.

IP_3 measurements can be made over a range of frequency spacings, with and without the front-end preamplifiers, and 2nd-order measurements can be made to assess front-end filter effectiveness. In-band linearity is assessed by setting the generators 200Hz apart, centred in the receiver passband and observing the audio output on a spectrum analyser. The resulting intermodulation products observed are a good indication of the linearity of the total signal path right through to the loudspeaker. It is quite noticeable that a really clean-sounding receiver generally shows intermodulation products some 40 to 50dB down on either of the two tones and with a less-clean-sounding receiver these levels are only 20 to 30dB down. The results usually hold over a wide range of signal levels and are often improved by reducing the RF gain control.

Sometimes slow AGC gives better results than fast. Front-end blocking is caused by gain compression in the receiver front-end stages ahead of the main IF filters. Generator 1 is set on-tune at a defined S-meter level (typically S9). Generator 2 is offset from the on-tune frequency and the level increased until the S-meter reading drops by 1dB. The level of generator 2 related to the receiver input is taken as the blocking level. Measurements at different offset frequencies are usually made. If reciprocal mixing is poor it may not be possible to measure blocking by this method, but then it is probably irrelevant as well. AGC is not normally applied to the front-end of modern receivers but, if it is, the measured blocking level will be dependent on the on-tune signal level. Measurement at lower on-tune levels may not be possible due to reciprocal mixing.

Receiver reciprocal mixing

This arises from phase noise on the receiver local oscillator mixing with a strong incoming signal and masking an adjacent weak signal.

Fig 7.26 shows the test arrangement for making reciprocal mixing measurements. It is most important that the sideband noise spectrum of the RF generator is considerably lower than that of the receiver local oscillator. In order to achieve this, measurements are made at a single frequency by inserting a narrow bandwidth crystal filter together with suitable matching components and attenuators between the generator and the receiver. For example, a 21.4MHz filter could be used and then all measurements relate to the 21MHz band. The generator should be tuned to the passband of the filter and as close to the roll-off as possible on the measurement side. The total loss of the filter, matching components, and attenuators are noted for calibration. The receiver is tuned away from the generator frequency, noting the generator level required to give a 3dB increase in noise output from the receiver. Hence the noise due to reciprocal mixing is then equal to the noise floor of the receiver.

The phase-noise-limited dynamic range is the difference between the generator level as seen at the receiver input and the receiver noise floor. Measurements are made in SSB bandwidths for compatibility with the other measurements, although CW bandwidths may really be better for close-in measurements. The phase noise of the receiver synthesiser in dBc/Hz can be calculated at the specified offset from the equation:

Oscillator noise (dBc/Hz) = - (PNDR+$10\log_{10}$B)

where PNDR is the phase noise limited dynamic range in dB, and B is the receiver noise bandwidth in Hz (typically 2500 for SSB).

8 Some Test Equipment and Ancillary Items that You Can Make

A SELECTION of useful reference articles to start:

→ Homebrew RF Test Equipment and Software

http://www.qsl.net/n9zia/wireless/appendixF.html

→ Amateur Radio Equipment schematics

http://www.one-electron.com/FC_Ham.html

→ *Arduino Projects for Amateur Radio*, Jack Purdum, W8TEE, Dennis Kidder, W6DQ, McGraw-Hill, 2015

→ *More Arduino Projects for Ham Radio*, Glen Popiel, KW5GP, ARRL, First Edition, April 2017 ISBN: 978-1-62595-070-3

→ MØPZT Amateur Radio - Arduino Projects

http://www.m0pzt.com/arduino/

→ PCB techniques for the design and layout of amplifiers, attenuators and sources:

'Homebrew Construction', Eamon Skelton, EI9GQ, *RadCom*, February 2016.

→ Useful tips on air- and ferrite-cored coils, and on HF/VHF construction techniques:

'Homebrew', Eamon Skelton, EI9GQ, *RadCom*, April 2016.

→ For guidance on fitting SMA connectors, see:

'Fitting SMA connectors', Mike Richards, G4NWC, *RadCom*, December 2016.

→ Having completed a project, you may want to fabricate a new scale for your analogue moving-coil meter. If so, visit the following site, which has two Windows programs: 'MeterBasic' and 'Meter', as well as lots of useful notes. The former is free, and will generate a simple scale; the latter is a more comprehensive, professional-level meter-scale drawing program, which requires a text-based key to operate, provided on purchase: *http://www.tonnesoftware.com/meter.html*

8.1 Probes

An RF Diode Probe

This allows a DC voltmeter to measure AC voltages into the tens of MHz, or even 100MHz, region. It simply rectifies the AC and then passes a DC voltage to the meter. The diode is often the limiting factor, as to get high speed operation the diode junction must be narrow, but this reduces its breakdown voltage. Using a BAT46 Schottky barrier diode, the maximum input voltage is about 35Vrms, whereas with a 1N914/1N4148/OA91 it is about 45Vrms. Using a Schottky diode, the forward voltage drop is of the order of 0.2 to 0.3V, but with a silicon type it is about 0.6V. The probe components should be mounted in a small metal cylinder for good screening,

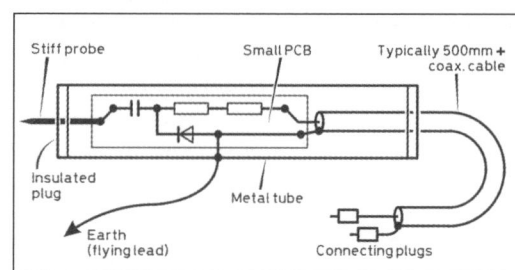

Fig 8.1 Typical construction of an RF probe

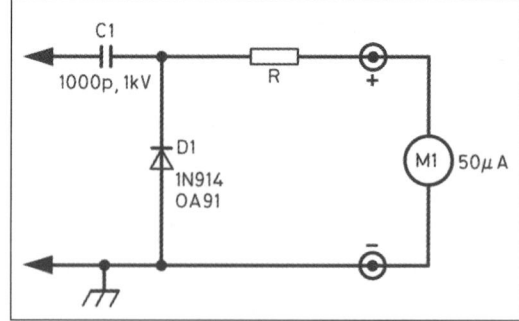

Fig 8.2 RF probe circuit. For R=270k+12k the meter scaling is 0-10V and full-scale power into 50Ω is 2W. For R=820k+27k the meter scaling is 0-30V and power in 50Ω is 18W

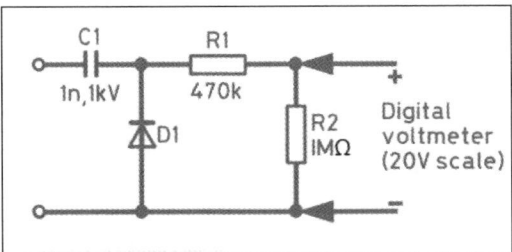

Fig 8.3 RF probe for digital voltmeter

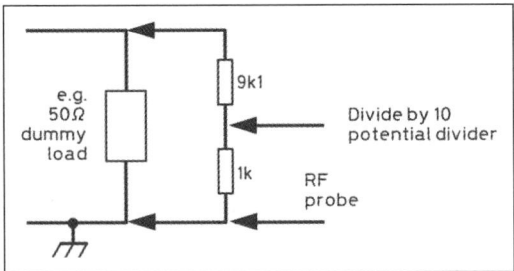

Fig 8.4 Suggested method for higher voltages

and the resulting DC signal fed via a coaxial cable to the DC meter - see **Fig 8.1**.

A typical circuit is shown in **Fig 8.2** with component values suitable for feeding a 50μA moving-coil meter. The advantage of arranging the capacitor and rectifier in this manner is that the capacitor also acts as a DC block.

To feed a high input impedance DC voltmeter (such as an electronic analogue meter or a digital meter) the alternative arrangement of **Fig 8.3** can be used. Here, the input resistance of the meter should be ten times the value of R2, and the series resistor R1 around 41% of the combined resistance of R2 in parallel with the meter input resistance. This allows the meter to read the RMS value of the RF signal. The values shown are suitable for a meter with an input resistance of at least 10MΩ. For a meter of input resistance of 1MΩ, reduce the values of R1 and R2 by a factor of 10.

To measure higher voltages and hence higher power levels, try a resistive divider across the load. **Fig 8.4** shows a divide-by-10 unit suitable for a 50Ω system. Remember the actual voltage is 10 times the meter reading!

An RF Millivolt Probe

Simple RF diode probes are limited to voltages in excess of about 1V as a result of the diode forward voltage drop. Some indication will be apparent below this, but the diode will be working in its 'square law' area. A method to extend measurements down to a few millivolts is to use the IC transistor array CA3046 which has a minimum gain-bandwidth product of 300MHz. The concept is to amplify the RF signal before detection. A suggested arrangement is shown in **Fig 8.5**. The device should be mounted in a small screened case with a probe for the RF input in the usual manner. Every effort should be made to keep stray capacitance to a minimum and no IC socket should be used. The input impedance should be about 50kΩ in parallel with 3pF. With two symmetrical DC Darlington pairs, the maximum offset voltage will be less than 1mV. The working range will be from about

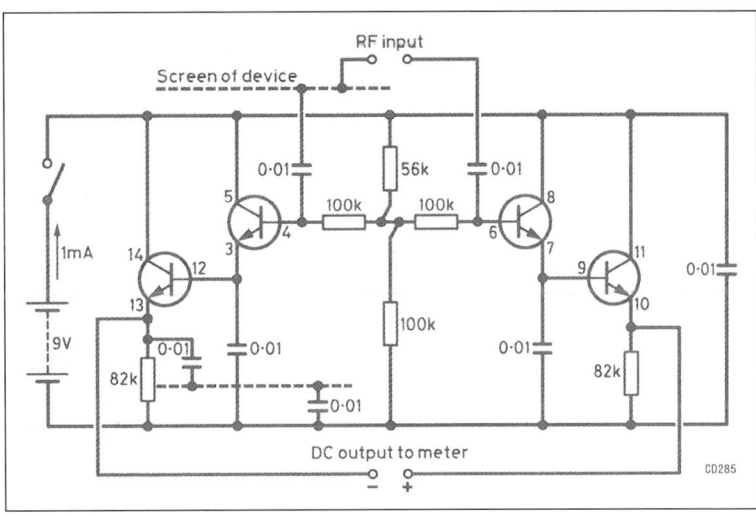

Fig 8.5 Peak reading RF voltmeter probe (Numbers refer to the CA3046 14-pin DIL pinout. All capacitors are disc type)

1mV to 4V and the device is intended as an add-on unit for a voltmeter. Calibration can be carried out in the several-volt region using a DC voltmeter. Useful measurements should be possible to frequencies in excess of 100MHz.

An E-Field Probe

A spectrum analyser normally expects to operate in a 50Ω environment whereas the location in the circuit where we need to test may well not be 50Ω. What is needed is a high impedance test probe to connect the circuit under test to the 50Ω input of the analyser. Commercial, very wide bandwidth FET probes can be expensive, even on the surplus market, but happily there is a simple solution that will work well to over 5GHz and costs very little to make. A 510Ω resistor, when connected at the input end of a suitable length of low-loss coaxial cable and to the 50Ω analyser input, will provide a -20dB probe of moderately high input impedance (>500Ω) as well as a very wide bandwidth - providing suitable parts are used. See the reference below for constructional details. **Fig 8.6** shows a possible form of probe construction. A simple fibreglass (FR4) PCB is etched with a 50Ω microstrip track down the centre. The track is cut to accommodate a 0805 size 510R surface mount resistor at one end next to the probe tip and the output extracted via a miniature SMA, SMB, or SMC RF connector of the type used for mounting at the edge of printed circuit boards. To this is then connected a suitable 50Ω coaxial cable.

A probe tip can be made by removing the silver- or gold-plated centre pin from an old

BNC connector and soldering this on to the pad at the input end of the 0805 resistor. No explicit earth lead is needed when the probe is used above a few hundred MHz as the stray capacitance between the probe and the circuit under test is usually sufficient.

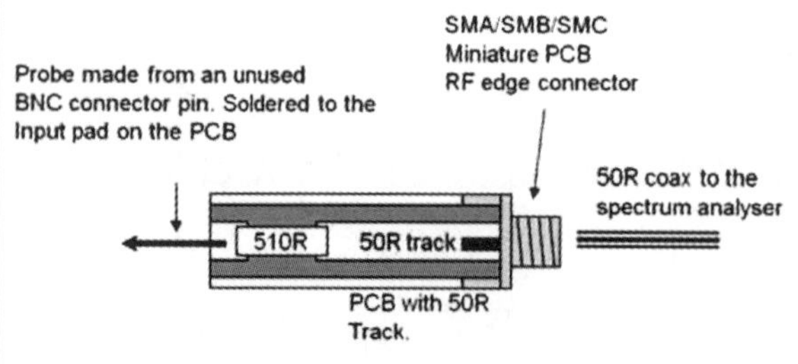

The probe should be housed within a short length of metal tubing, and when working correctly, the PCB can be fixed with a two-part adhesive such as Araldite. Try to ensure a good internal ground connection between the tube and the PCB ground; this can be provided using a short length of phosphor-bronze strip soldered to the PCB ground and bent to contact the tubing. No DC blocking capacitor is required if the spectrum analyser already has a DC blocked input, otherwise a 100pF 0805 size capacitor can be connected in series with the resistor.

→ See 'The GHz bands', by Sam Jewel, G4DDK, *RadCom*, June 2008, p38.

An H-Field Probe

Previously, a simple E-field (electric) probe was described that could be used for signal tracing when used with, for example, a spectrum analyser. An equally useful device is the H-field (magnetic) probe for general signal tracing, and especially for identifying sources of RF noise. The H-field probe is particularly good for the latter as the H-field decreases rapidly with distance from the source. This enables a problem area of, say, a PCB to be localised very quickly.

Since the sensitivity of the probe loop depends on its diameter, smaller loops will be less sensitive, but allow easier localisation of the source, as the probe needs to be closer to the source to pick up sufficient signal. It is worth making probes with a selection of loop diameters from a few millimetres to about 50mm. The bigger the diameter of the loop, the easier it is to make.

H-field probes can be made in many different ways. The one shown in **Pic 8.1** uses a short length of semi-rigid coaxial cable, such as UT141 or

Fig 8.6 Construction of a high impedance probe for microwave signal tracing

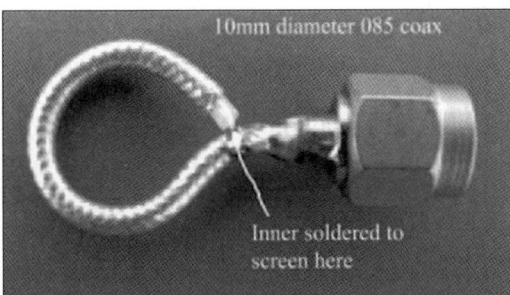

Pic 8.1. A magnetic field probe. This probe uses QuickForm 085 coax and has been found useful to over 5GHz. Note how the coaxial cable inner is soldered to the screen at the end of the loop. The screen is left unconnected at this point

UT085. By using coaxial cable, the probe loop can be shielded to reduce any unwanted E-field pick up.

Suitable semi-rigid cable can often be found at radio rallies and may already have an SMA, SMB or SMC connector on both ends. Only one connector is needed, so the cable may provide one with enough material to make two probes but of different diameters.

For a general purpose probe, cut the cable to about 50-70mm long. Remove 1.5mm of outer jacket from the cable, exposing the PTFE insulation over the inner. Remove 1mm of the PTFE to expose the coaxial cable inner. Bend the coaxial cable around a suitable former of approximately 10mm diameter, being careful not to kink it. Other sizes of former can be used, but 10mm has proven to be a good size for most purposes. Bending the cable so that it doesn't break or kink can be difficult and one can waste several pieces of cable before it is correct. For this reason flexible 'QuickForm' semi-rigid cable is probably a better choice; UT141 size cable is probably the best provided that it is not bent any tighter than about 15mm diameter

Once the loop is formed, solder the exposed inner of the cable to the outer (screen) of the loop, as shown in **Pic 8.1**. Very small loops can be made by simply connecting the extended inner di-

rectly to the coaxial cable outer. Do not make the loop too small or the probe will be very insensitive. The loop should preferably be at least 2-4mm in diameter.

Since the loop screen is effectively connected to ground, be careful not to short it to the circuit under test - it may be advisable to cover the loop with suitable insulation. Very small loops can be 'potted' in a two-part adhesive such as Araldite.

→ See 'The GHz bands', by Sam Jewel, G4DDK, *RadCom*, June 2008, p40.

A Discharge Probe

The need for, and use of, a discharge probe, also known as a 'grounding stick' or 'grounding hook', is described in Chapter 6.5. **Fig 8.7** shows how to construct such a probe of your own. A large screw eye can be substituted for the hook. It is important that the grounding wire be left out-side the handle, so that any hazardous voltages or currents will not be present near your hand or body. Some authorities recommend including a series resistance to restrict the current so that electrolytic capacitors are not discharged too quickly.

→ *The ARRL Handbook 2010*, 87th Edn, Chapter 7, ARRL, 2010.

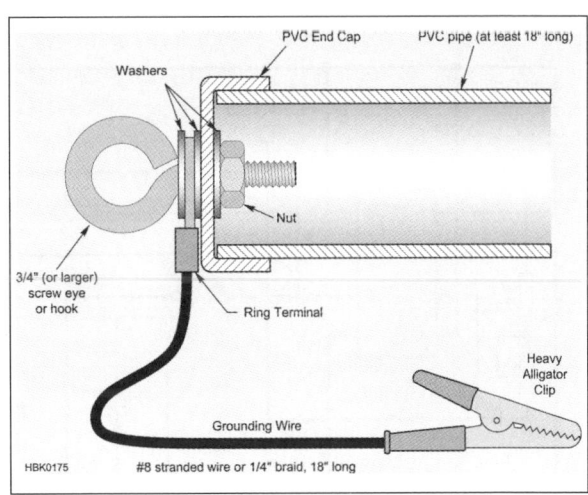

Fig 8.7 Example of a grounding probe to discharge capacitor energy safely

8.2 LF Signal Sources

Audio Sources

Apps that can generate both fixed and variable tones can be downloaded quickly and easily for Android and OS mobile phones and tablets – just search on 'tone generator'.

A Windows 10 compatible tone generator is available from the website below.

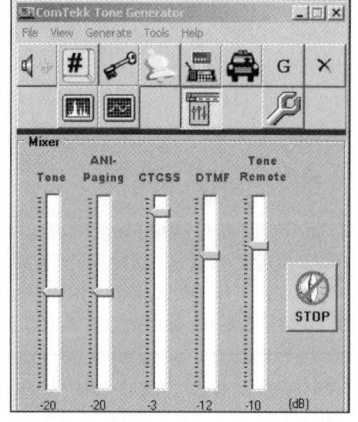

Pic 8.2 ComTekk tone generator

In the trial version, only Tone and CTCSS are active. It allows DTMF, CTCSS, paging and other tones to be generated for testing radio equipment – see **Pic 8.2**

The software allows auto-calibration and has an audio oscilloscope and spectrum analyser.

http://comtekk.us/tone-generator.htm

Square-wave generator 10Hz-1MHz

This oscillator provides a square-wave output at 5V and 12V amplitudes which makes it suitable for testing both TTL and CMOS circuits. The switched ranges produced are approximately as follows (slight variations can be expected due to component tolerances).

Range 1   9Hz - 110Hz
Range 2   91Hz - 1.1kHz
Range 3   910Hz - 11kHz
Range 4   9kHz - 110kHz
Range 5   90kHz - 1MHz

Circuit description

The circuit is straightforward and is based on a low power 555 timer, the TLC555M (IC1). R1 and RV1, together with capacitors C3 to C7, are the frequency determining components, switch SW1 selecting the capacitor for the various ranges. Note that the capacitor for the highest frequency ranges is only 47pF; this takes into account stray capacitance in the circuit and timing restraints within the IC. The outputs are a fixed 5V or 12V square-wave from buffers TR2 and TR1 respectively. Resistors R7 and R8 define the outputs at approximately 50Ω. Voltage regulator IC2 is required to produce the 5V supply for TR2. The circuit diagram is shown in

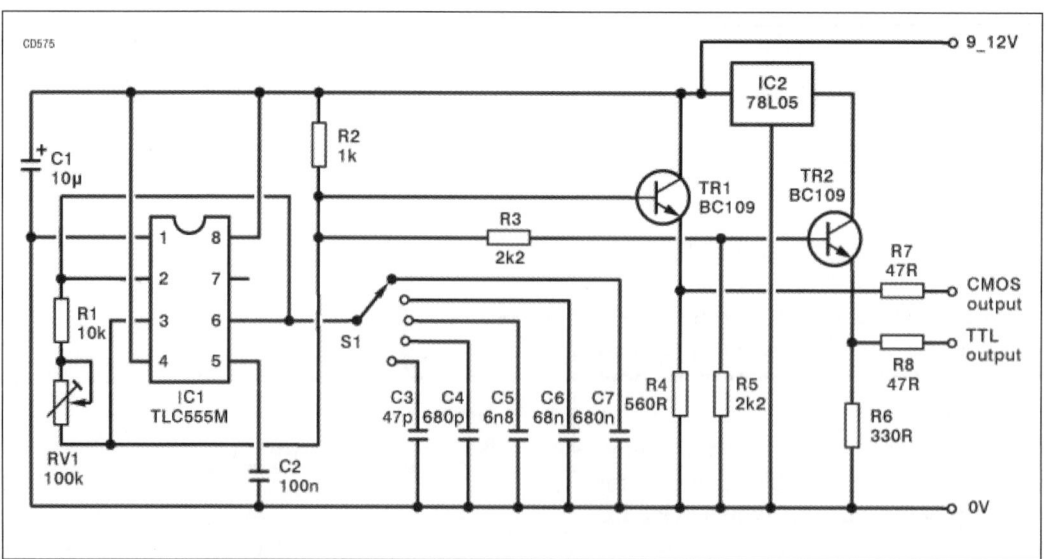

Fig 8.8 Square-wave generator

Fig 8.8. Easy to construct, a PCB layout and component overlay is given in Appendix D. The component list is in **Table 8.1**.

R1	10k	C4	680p, polystyrene, 2.5%
R2	1k		
R3	2k2		
R4	560R	C5	6n8, polystyrene, 2.5%
R5	2k2		
R6	330R	C6	68n, polyester, 7.5mm pitch, 5%
R7,8	47R		
RV1	100k log pot.	C7	680n, polyester, 7.5mm pitch, 5%
TR1,2	BC109 or similar		
C1	10µ, 25V	IC1	TLC555M or equivalent
C2	100n, ceramic		
C3	47p, polystyrene, 2.5%	IC2	78L05, 100mA, 5V regulator
		SW1	6p, 2w rotary switch

Resistors are 0.25W/0.5W, 5% unless specified otherwise.

Table 8.1 Component list for a square-wave generator

Wideband signal injector

This is a very useful aid for servicing equipment from audio frequencies up to radio frequencies. It enables a signal to be injected at some point via the probe and the response either listened for, or searched for with an oscilloscope. This unit generates a rectangular wave at about 2.7kHz with a mark to space ratio of about 2:1. This generates sufficient harmonics to be useful at low RF.

Circuit description

The circuit is based around a low power 555 timer - a TLC555, see **Fig 8.9**. Resistors R1 and R2 determine the mark to space ratio, and in conjunction with C2, the frequency of oscillation. The rectangular-wave output is via resistor R4 (which limits the current), and capacitor C4 is for DC isolation. A component list is given in **Table 8.2**.

Construction

The wideband injector circuit should be housed in a metal container if possible together with a battery (possibly a PP3) and an on/off switch (**Fig 8.10**). If an external supply is

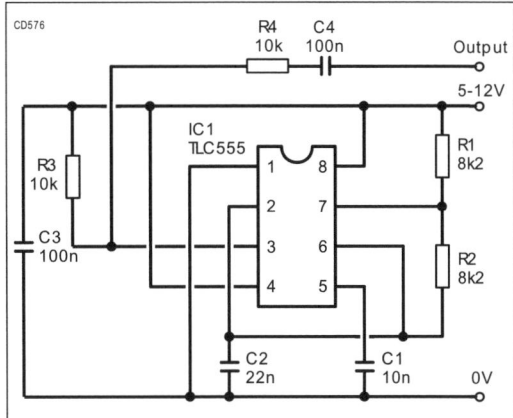

Fig 8.9 Wide-band signal injector

R1,2	8k2	IC1	TLC555 or equivalent
R3,4	10k		
C1	10n	Resistors are 0.25W/0.5W, 5%	
C2	22n		
C3,4	100n		

Table 8.2 Component list for wide-band signal injector

used, power switching can also be external. The probe should ideally be sleeved so that there is no inadvertent touching and damaging of circuit components. A flying lead with crocodile-clip or similar is required to make the ground connection, although at RF frequencies this may be found unnecessary.

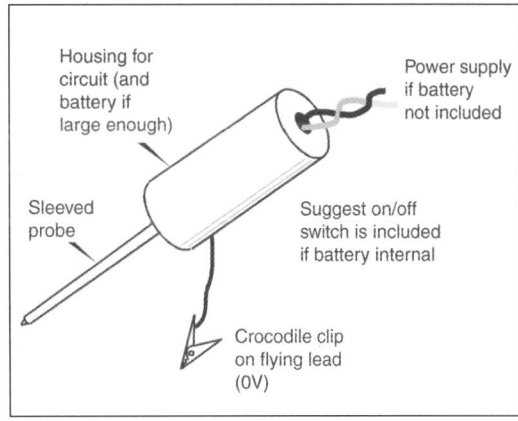

Fig 8.10 Typical construction for wide-band signal injector

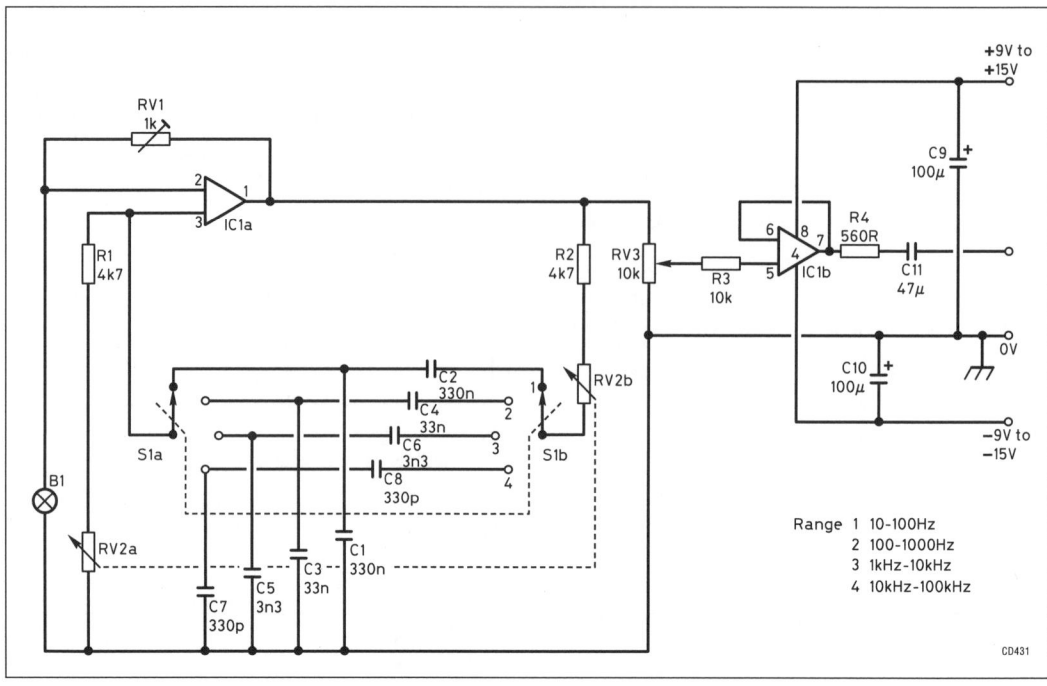

Fig 8.11 Circuit diagram for low-frequency oscillator

Sinewave oscillator 10Hz-100kHz

This circuit covers the above frequency-band in four switched ranges, and requires a symmetrical plus-and-minus supply between 9 and 15V.

Circuit description

The circuit diagram for this oscillator is shown in **Fig 8.11**. It is based on a Wien-bridge oscillator formed around IC1a and buffered by IC1b. The main frequency determining components are R1/R2 and RV2 with capacitors C1 to C8. In the configuration shown, stable oscillation can occur only if the loop gain remains at unity at the oscillation frequency. The circuit achieves this control by using the positive temperature coefficient of a small lamp to regulate the gain as the oscillator varies its output. Potentiometer RV3 forms the output level control with R4 giving a defined output resistance of approximately 600Ω with C11 providing DC isolation. Capacitors C9 and C10 provide power supply line decoupling. See **Table 8.3** for the component list.

R1,2 4k7	C11 47µ, bipolar
R3 10k	RV1 1k trimmer
R4 560R	RV2 47k dual gang pot.
C1,2 330n	
C3,4 33n	RV3 10k lin. pot.
C5,6 3n3	B1 28V, 40mA bulb
C7,8 330p	IC1 LM358
C9,10 100µ, 25V	
Resistors are 0.25W/0.5W, 5% unless specified otherwise.	

Table 8.3 Component list for the LF sine-wave oscillator

Construction

The layout of the circuit is not critical, but to help, a PCB layout is given in Appendix D. If some ranges, or the output level control, are not required, the layout can be tailored accordingly. The feedback resistor RV1 should be adjusted so that the output on all ranges is just below the clipping level.

Testing

No frequency calibration is required of this circuit, though it would be wise to check with a frequency counter that the ranges are as suggested. An oscilloscope is required for setting up the adjustment of RV1.

Two-tone oscillator

This could be constructed by building two of the previous oscillators and then combining their outputs. The outputs should be set to two non-harmonically related frequencies such as 1.4kHz and 2kHz or 1kHz and 1.8kHz.

An alternative approach is to build two fixed-frequency oscillators, and combine their output, possibly via a switch, so that each tone can be used independently. This will provide a self-contained unit for producing two tones for the testing of sideband transmitters. The frequencies produced are non-harmonically related at approximately 1.4kHz and 2kHz. The circuit was originally introduced in the second edition of this book

and modified to also give a burst output in the third edition. Unfortunately the MC3340 (IC2) is discontinued although it may occasionally be found on the surplus market and/ or Internet auction sites. A straight two-tone oscillator can still be made by omitting IC2 and IC3 and associated components. Connect C14 (pin 1 end) to the top end of RV1.

Circuit description

The two oscillators are formed around quad op-amp IC1 - see **Fig 8.12**, and **Table 8.4** for the associated component listing. The sine-wave outputs from these are routed via SW1, which allows either one tone through, or both together. The resulting signal is summed by resistors R13 and R14 before being fed into an electronic attenuator, IC2. The attenuator gain is controlled by a DC voltage on pin 2. For the condition of continuous output this pin is connected to -6V via R15.

For the burst mode the square-wave output from IC3 is fed into the integrating network formed by R15/C13. This modified signal controls the gain of IC2 and provides a burst

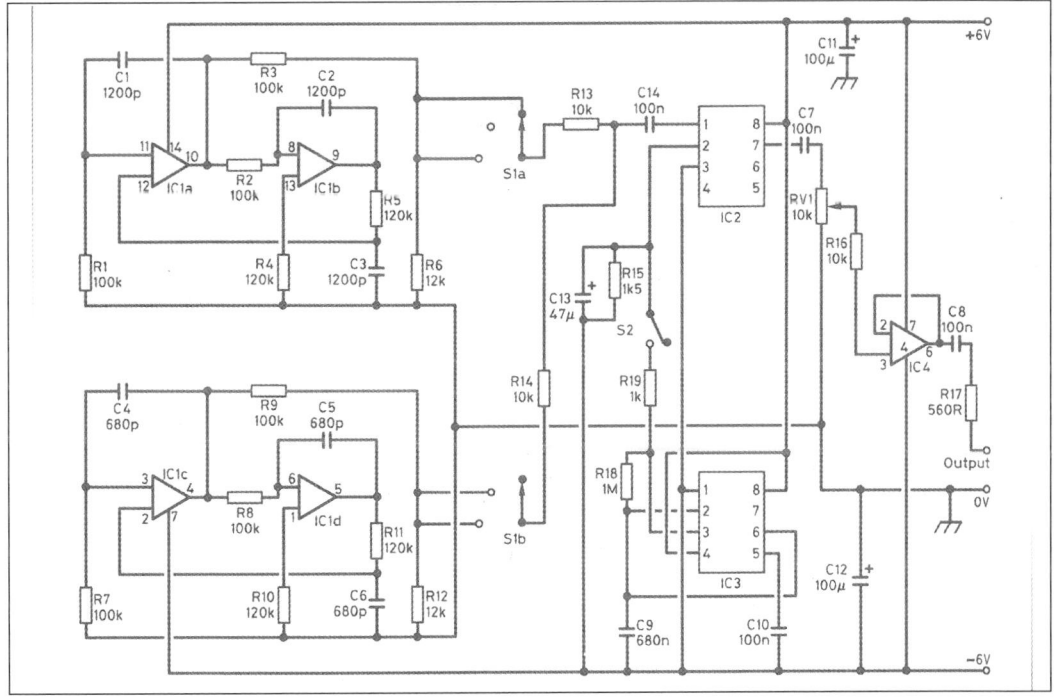

Fig 8.12 Circuit diagram of two-tone oscillator

R1,2	100k	R19	1k	IC1	LM3900
R3	100k	RV1	10k lin pot.	IC2	MC3340P
R4,5	120k	C1,2,3	1200p	IC3	TLC555M
R6	12k		polystyrene	IC4	741
R7,8	100k	C4,5,6	680p	SW1	4p, 3w rotary
R9	100k		polystyrene		switch
R10,11	120k	C7,8	100n	SW2	SPST PCB
R12	12k	C9	680n		mounting
R13,14,16	10k	C10	100n ceramic		
R15	1k5	C11,12	100µF		
R17	560R	C13	47µF, 16V		
R18	1M	C14	100n, ceramic		

Resistors are 0.25W/0.5W, 5% unless specified otherwise.

Table 8.4 Component listing for two-tone oscillator

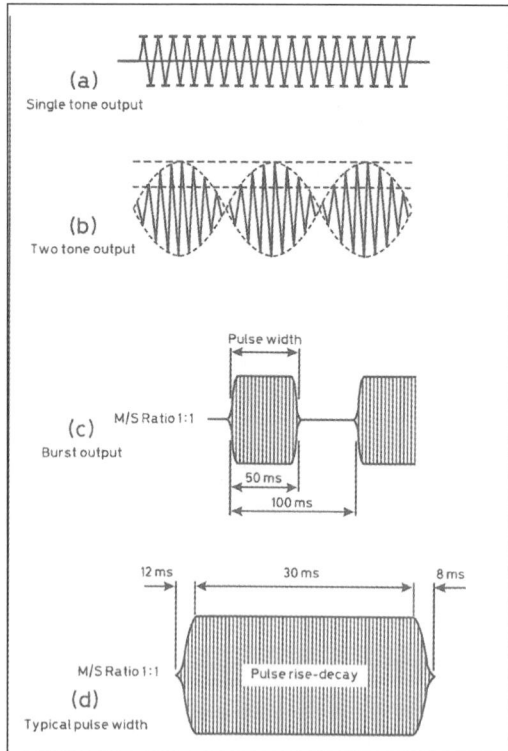

(a) Single tone output

(b) Two tone output

(c) Burst output — M/S Ratio 1:1 — Pulse width — 50 ms — 100 ms

(d) Typical pulse width — M/S Ratio 1:1 — 12 ms — 30 ms — 8 ms — Pulse rise-decay

Fig 8.13 Typical output waveforms of two-tone oscillator

Construction

Construction is straightforward and requires no special techniques. The burst option can be omitted if required, but ensure that pin 2 of IC2 is connected to -6V via R15. C13 can be omitted in this case as it only shapes the envelope of the pulsed output. A PCB and component layout is given in Appendix D.

Testing

No calibration of the circuit is necessary but the operation should be checked. **Fig 8.13** gives typical outputs for the tone output, continuous or pulsed.

8.3 RF Signal Sources

A few circuit suggestions from HF to microwave.

HF Variable Frequency Source

The circuit in **Fig 8.14** will tune in excess of 500kHz in the vicinity of the 80m band using the values shown. However, by modifying the values of L1, C1 and C2, the tuning range can be modified to cover a suitable segment of any part of the HF bands.

In general, slug tuning of a suitable coil will set the band centre, and the value of the two capacitors will set the tuning range. The capacitive divider (which controls feedback) may have to be reduced in effective value for

output as shown in Fig 12.6. The output from the attenuator is buffered by IC4, the output level control being set by RV1. The output resistance is defined by R17 at approximately 600Ω.

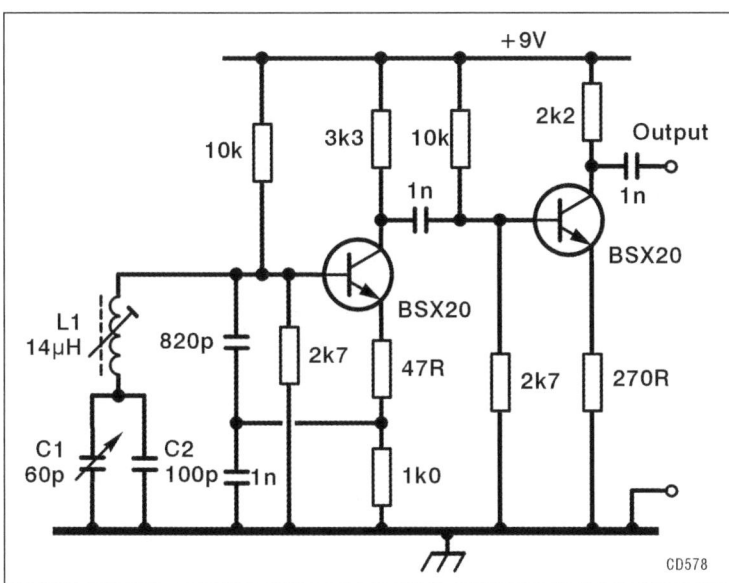

Fig 8.14 An HF bands signal source

signal up to 30MHz and 50MHz respectively. Such a module can be controlled from an Arduino processor board, which itself costs less than £20 and can be programmed using their open-source platform. This gives great flexibility at low cost. As with any synthesised source, close-to-carrier noise is likely to be quite poor compared to a non-synthesised source, but this may not matter in some applications.

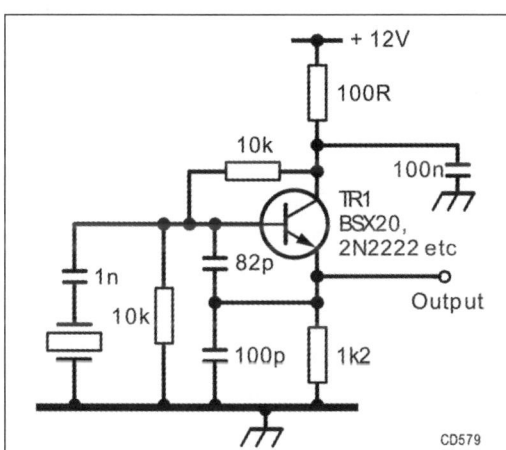

Fig 8.15 Crystal oscillator for 1 - 20MHz

the higher frequencies. By switching several inductance values and maybe several values for C2, it would be possible to make a generator covering a number of bands. However, the output which is of the order of tens of millivolts would vary with frequency.

HF Synthesised Source

Direct Digital Synthesis (DDS) modules such as the AD9850 and AD9851 can be purchased ready-made for just a few pounds, and are able to generate a sinewave RF

HF Crystal Controlled Source

Fig 8.15 shows a crystal oscillator running in fundamental mode (usually 1 to 20MHz) with typical circuit values. Some adjustment may be required as the frequency rises. Above about 20MHz, crystals function on their overtones and so a tuned circuit must be provided in the output of the active device to ensure that the correct overtone is selected.

VHF Crystal Controlled Source

Fig 8.16 shows two versions of a crystal oscillator using 8MHz-range crystals. The output should be checked with an absorption wavemeter to make sure that the correct harmonic of the crystal frequency has been selected. The possibility of error would be reduced if a higher frequency crystal were used; this would be particularly desirable if the output circuit were pushed to give an output on 432MHz.

In **Fig 8.16**(a) L1 is 20t, 8mm diameter; L2 120mm long by 3mm diameter; L3 coupling loop 16SWG wire; RFC is 35t, 6mm diameter. The inductor details for **Fig 8.16**(b) are L1 28t 28SWG wire, 6.3mm diameter; L2 is 4t 18SWG wire, 12.5mm diameter and L3 coupling loop 1t 18SWG wire.

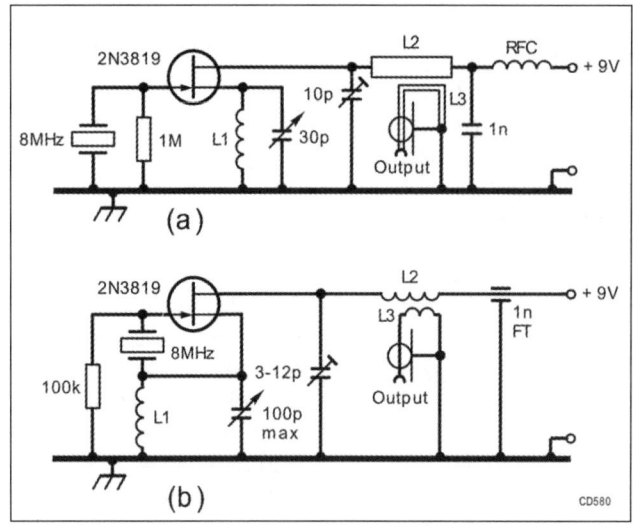

Fig 8.16 Two alternative VHF signal sources

→ See also, how to make a 144MHz oscillator using a 16MHz Colpitts crystal oscillator/tripler & tripler: 'Homebrew', Eamon Skelton, EI9GQ, *RadCom*, April 2016.

UHF and above Signal Sources:

Construction of a useful 23cm signal source (as well as a mixer and band-pass filter) is described in 'Homebrew' by Eamon Skelton, EI9GQ, *RadCom*, June 2016.

→ Visit *www.G4DDK.com* for useful information and links.

10GHz Crystal Controlled Frequency Marker

This unit generates a large number of signals at precisely known frequencies which can be used to accurately calibrate wideband receivers and transmitters.

Although the output power of individual harmonics is very low, signals are detectable with efficient receivers even with 40 to 50dB of attenuation between the unit and the receiver. This means that the unit can provide a rough check on the sensitivity of a receiver; signals should be detectable with even the most insensitive receiver. The output power is more than adequate for use as a frequency reference for an AFC system, but insufficient for

the unit to be employed as a signal source for tuning antennas. The range of the unit is only about 10m when antennas of 15dB gain are used.

The unit produces relatively strong signals at 96MHz spacing, with signals about 20dB weaker spaced every 48MHz. The choice of 48 and 96MHz as stage frequencies represents a compromise between generating a reasonable number of signals within the tuning range of most receivers while minimising the risk of confusion due to difficulty in identifying each of the harmonics. Other crystals may, of course, be used but there is an obvious advantage in using 'round number' frequencies and especially such frequencies as 36, 54 and 72MHz which also produce harmonics at other amateur frequencies and are therefore more easily measured.

Note that no provision is made for modulating the output. This is unnecessary if the local oscillator of the receiver can be frequency modulated with a tone which greatly assists finding signals. In calibrating a receiver it is essential that its local oscillator is already calibrated to within 20MHz using a wavemeter.

Circuit description

The circuit diagram of the unit is shown in **Fig 8.17** and its component list in **Table 8.5**. TR1 forms an oscillator on 48MHz with XL1. This is followed by a doubler formed by TR2. The resulting output at 96MHz is fed to a mixer diode, D1, which is used as the final multiplier. TR2 should be fitted with a small heatsink and its emitter lead kept as short as possible, typically 3-4mm. The output of TR2 is fed via a bandpass filter to the mixer diode in order to reduce 48MHz feed-through.

Constructional details

The VHF circuitry is mounted on a piece of single-sided printed circuit board bolted to the inside of the lid of a 111x60x31mm (or similar) die-cast box (**Fig 8.18**).

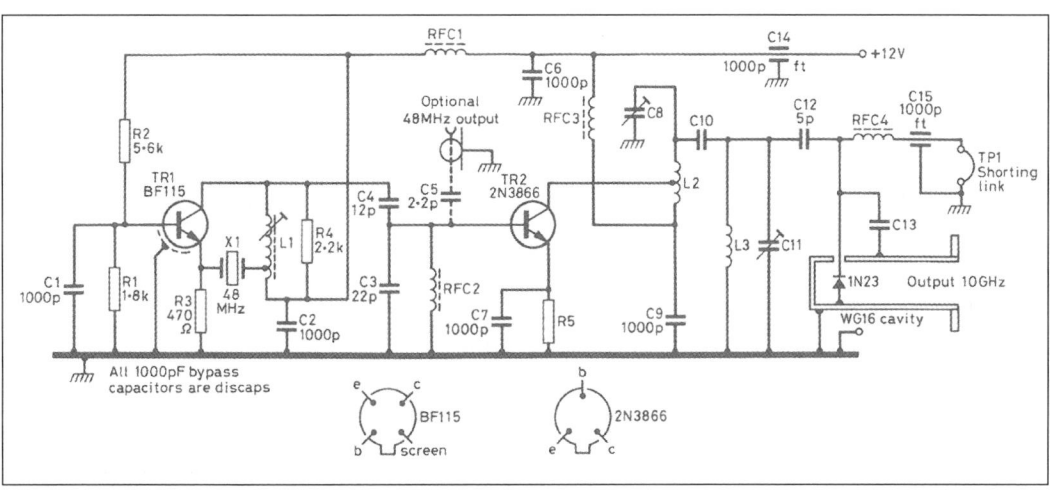

Fig 8.17 Circuit of the 10GHz frequency marker

R1	1k8	TR2	2N3866		1T from C2 end
R2	5k6	XL1	48MHz third overtone	L2,3	5T, 28swg tinned
R3	470		crystal		copper wire, 6mm
R4	2k2	C10	Two lengths of		inside dia, 12mm
R5	68-100* see text		thin single strand		long, L2 is centre
C1,2	1n, ceramic disc		insulated wire twisted		tapped
C3	22p, ceramic disc		together for 12mm	L4	2.5T,28swg ECW on
C4	12p, ceramic disc	C12	Formed by a 0.005in		one FX1115
C5	2p2, ceramic disc		PTFE or polythene	L5,6	As L4 but on two
C6,7	1n, ceramic disc		sheet between the		FX1115
C8,13	1n, feed-through		end of the 1N23	L7	10T, 28swg on one
C9	1n disc ceramic		diode and the		FX1115.
C11	4p7, ceramic disc		waveguide wall		
D1	1N23	L1	10T, 28swg ECW on		
TR1	BF115		8mm former, tapped		

WG16   60mm minimum length waveguide 16

Resistors are 0.25W/0.5W. 5% unless specified otherwise

Table 8.5 Component list for 10GHz frequency marker

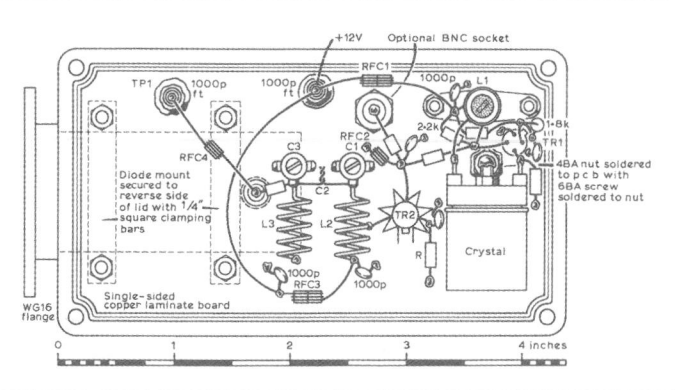

Fig 8.18 Component layout of the 10GHz frequency marker

The final multiplier (**Fig 8.19**) consists of a length of WG16 waveguide at least 60mm long, closed at one end, with the mixer diode D1 (1N23) mounted centrally and 7.5mm from the closed end. The waveguide is clamped to the outside of the lid, connection to the diode being made via the inner of a Belling Lee TV socket which passes through holes drilled in the printed circuit board and the lid.

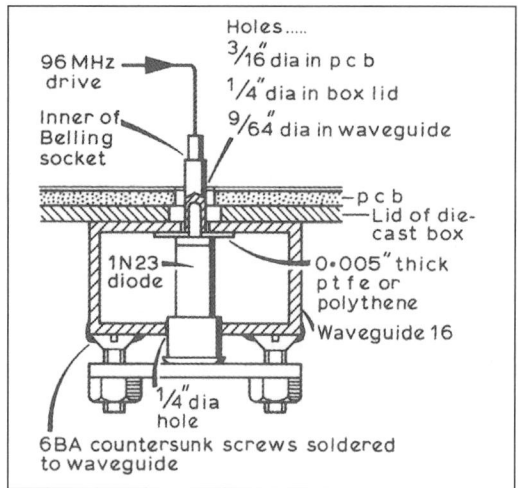

Fig 8.19 Section showing construction of the multiplier signal source

Alignment

Remove the link LK1 and replace with a milliammeter (50mA FSD), this will measure the diode current. In aligning the unit, L1 is adjusted to produce the maximum voltage across R5; C1 and C3 are adjusted and re-adjusted to maximise the diode current. The value of R5 is then changed as necessary to set the diode current to 20-25mA. The frequency of the crystal oscillator can be checked on a counter via the optional output connector or using, for example, a 144MHz receiver. Maximum output may not coincide exactly with maximum diode current.

Instability may occur at maximum diode current, so it is worth checking with a 144MHz and 10GHz receiver that the unit is working correctly. Some useful performance parameters are:

- Voltage supply: 12V DC
- Current required: 25-40mA
- Frequency pushing: 8kHz/V
- RF level of 96MHz harmonics: 10.0-10.2GHz: -75dBm
- RF level of 48MHz harmonics: 10.0-10.2GHz: -95dBm

Using the frequency marker

The marker is preferably connected direct-ly to the receiver via a variable attenuator, but alternatively it can be spaced from the receiver input by a few metres. With 20-30dB attenuation, the receiver should detect weak signals which are harmonics of 96MHz only. If the attenuation is reduced to around 10dB, the 48MHz intermediate harmonics should be heard as weak signals with the now strong 96MHz harmonics.

It is possible that more signals than expected will be heard as a result of the receiver having poor or even no image rejection. Thus a receiver with a 30MHz IF will respond to relatively strong harmonics at 9.984GHz (104x96MHz) when its local oscillator is tuned to either 9.954 or 10.014GHz and to signals at 10.080GHz (105x96MHz) when tuned to either 10.050 or 10.110GHz.

For the same reason, the receiver may also respond to the weaker harmonics at 48MHz spacing. The receiver local oscillator may therefore be calibrated precisely at several points, from which the (two) corresponding signal channels may be determined, provided the IF is known accurately. If a larger number of calibration points is required, either the IF may be temporarily changed, or a different crystal used in the marker.

It is possible to use the unit 'in the field', since it is small enough to be waved about in front of the receiver prior to establishing a contact. Alternatively the unit could be built into the system, coupled to the receiver through a 10dB directional coupler, to provide an instant check on its calibration and, in turn, that of the transmitter.

It is worth noting that the 10GHz multiplier stage can be replaced with one for any of the lower microwave bands, hence producing a unit capable of providing calibration signals on any of the intermediate bands.

8.4 Frequency Measurement

Frequency Counter & UHF prescaler

→ For construction of a 50MHz counter see 'Make it count', by Eamon Skelton, EI9GQ, in *RadCom*, October 2006 and his complimentary VHF/UHF prescaler in *RadCom* Jan-Feb 2008. Visit his support page at: *http://home-page.eircom.net/~ei9gq/counter.html*

GPS controlled frequency standard

→ see *www.g4jnt.com/freqlock.htm*

Simple Absorption Wavemeter (65-230MHz)

The absorption wavemeter shown in **Fig 8.20** is an easily built unit covering 65-230MHz.

Construction is straightforward and all the components, apart from the meter, are mounted on a perspex plate of thickness 3-4mm and measuring 190x75mm. Details of the tuned circuit are shown on **Fig 8.21**(a) and should be closely followed. The layout of the other components is not critical provided they are kept away from the inductor. RFC1

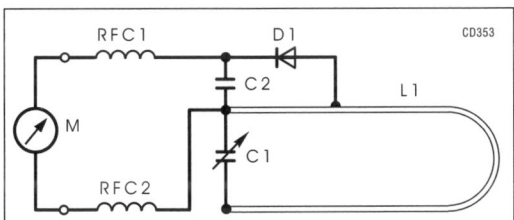

Fig 8.20 Circuit diagram of simple absorption wavemeter for 65-230MHz

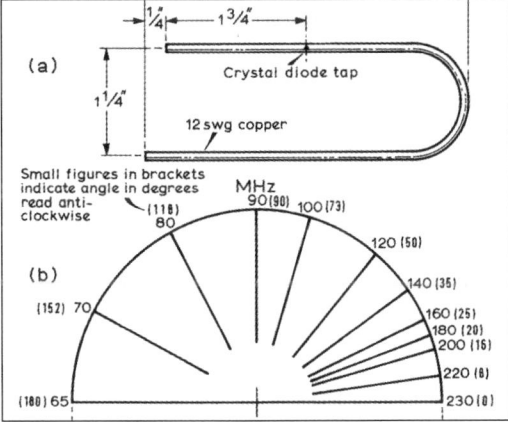

Fig 8.21 Constructional details of simple absorption wavemeter showing (a) inductor L1 and (b) dial plate. Metric equivalents are ¼in=6.3mm, 1¼in=31.8mm, 1¾in=44.5mm, 3in=76.2mm and 4in=101.6mm

and 2 are made by winding 80T of 40SWG ECW on a 10k, 0.5W resistor; D1 is an OA91, BAT85 or equivalent; C1 is a 4-50pF variable and C2 a 470pF ceramic capacitor.

For accurate calibration a signal generator is required but, provided the inductance loop is carefully constructed and the knob and scale are non-metallic, the dial markings can be determined from **Fig 8.21**(b).

In operation, the unit should be loosely coupled to the circuit under test and the capacitor tuned until the meter indicates resonance (a maximum). For low power oscillators etc a more sensitive meter should be used (eg 50µA or 100µA).

The wavemeter can also be used as a field strength indicator when making adjustments to VHF antennas. A single turn coil should be loosely coupled to the wavemeter loop and connected via a low impedance feeder to a dipole directed towards the antenna under test.

Through-Line Wavemeter (125-350MHz)

This type is intended for connection in the co-axial cable between an exciter or amplifier, and the antenna feeder. In the latter case however, the wavemeter output may need to be terminated with a 50Ω load and its input fed from a T- or coupler inserted between the amplifier and the antenna feeder to prevent damage to the wavemeter from the higher power.

The main element of this type of wavemeter is a short section of coaxial line within the unit to which is coupled the tuned circuit. Two methods of coupling are offered – *indirect* and *direct*. The dimensions for both types are given for fitting into a die-cast box which is 60mm wide externally.

Indirect coupling method

Here the tuned circuit is coupled to the inner of the coaxial line through an intermediate loop (**Fig 8.22**). This loop consists of a short length of rigid wire (coupled to the tuned circuit) which is connected at each end to a fine wire inserted into the coaxial line (and there-

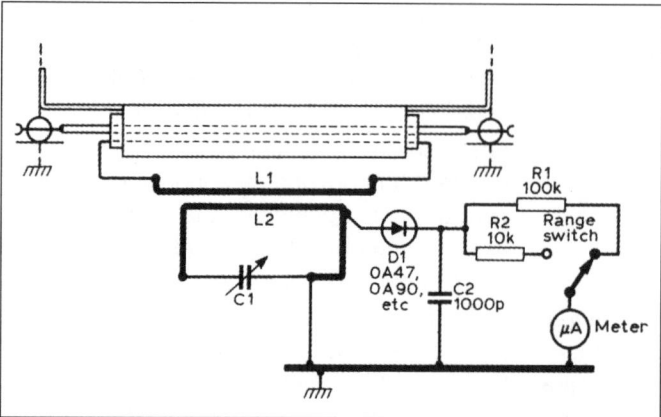

Fig 8.22 Schematic of the indirectly coupled through-line wavemeter

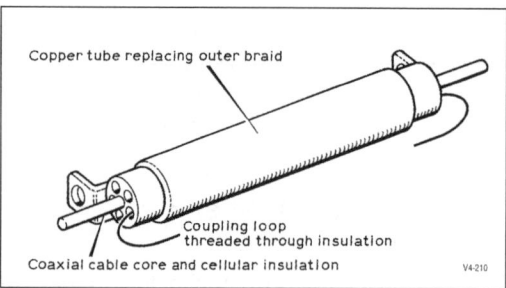

Fig 8.23 Indirect coupler construction

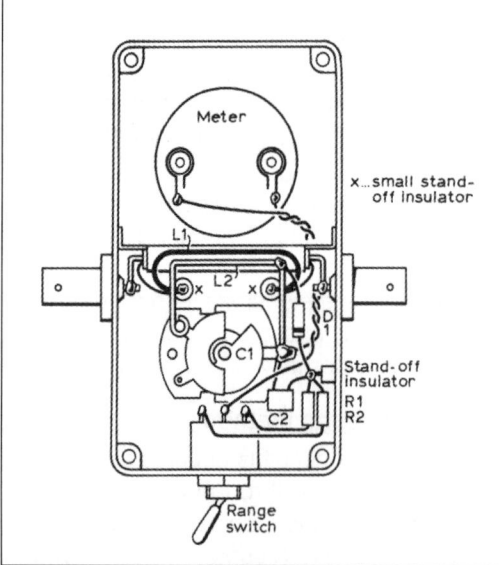

Fig 8.24 Arrangement of components for an indirectly coupled through-line wavemeter

fore couples to the field in the coaxial line).

When laying the coaxial line within the unit, it will be found convenient to fabricate this from a short piece of semi-air-spaced cable, replacing the outer braid with a short length of copper or brass tube of the same diameter in order to maintain the correct impedance - see **Fig 8.23**.

The cable used in the proto-type had an insulation diameter of 6.35mm (¼in) which was replaced by a copper tube. To change the impedance from 50Ω to 75Ω, simply replace the inner conductor with a thinner wire; the size can be either calculated or obtained from various Internet sites.

Fitting the coaxial element into the case (**Fig 8.24**) requires the ends of the tube to be shaped and bent so that the lugs thus produced can be fixed to the box by the screw holding the connector. In the design shown, BNC sockets have been used. The component arrangement is clearly shown and should present no problems. The connection of the diode should be 20mm from the ground-end of the tuned loop.

Direct coupling method

Here the intermediate coupling loop is avoided, and there is direct coupling of the tuned circuit to the inner conductor through a slot in the coaxial-outer.

Details of a suitable slotted line are given in **Fig 8.25**. The tuned circuit inductance consists of a simple 'U' shaped piece of 18SWG (1.25mm) enamelled copper wire to which the detector diode is connected at a point 20mm from the earth end. Spacing the loop from the inner conductor is important and should be within the field of the outer of the coaxial line. To aid alignment, the slot is arranged to be vertical, and this means that the fixing lugs of the tube forming the outer of the line must be at 45° to the slot to enable the

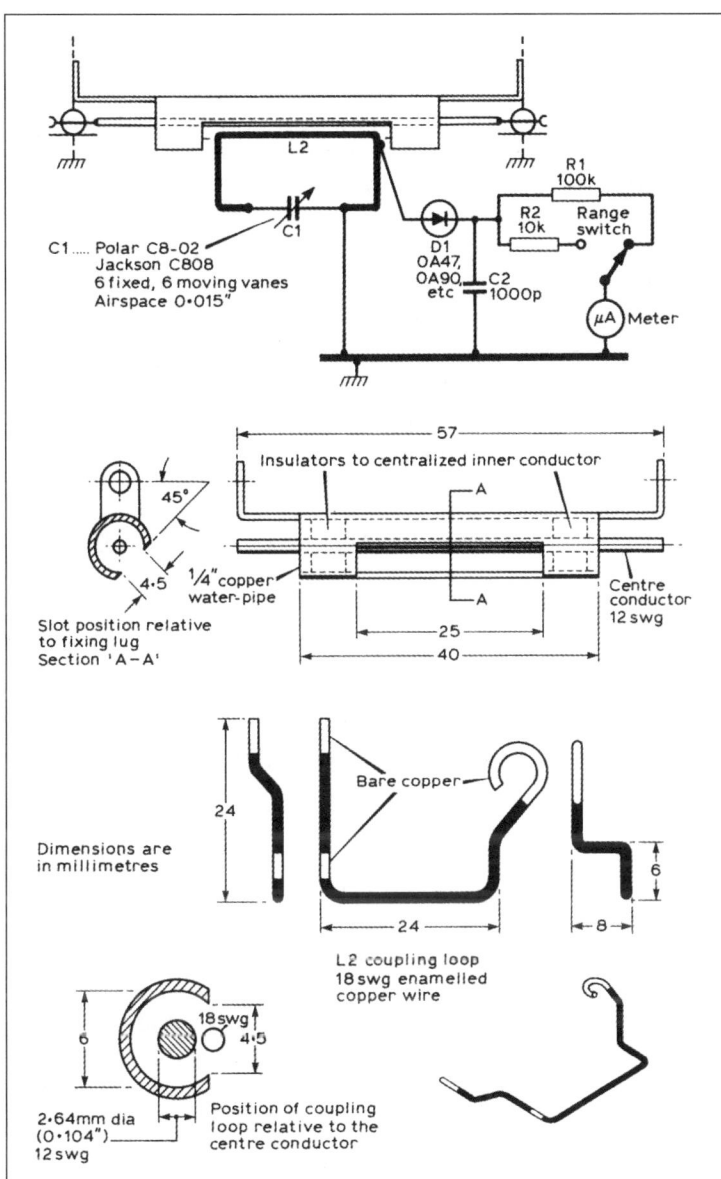

Fig 8.25 Schematic of a directly-coupled through-line wavemeter

8.5 Signal-to-Noise Optimiser and Interference Locators

Receiver Alignment Aid

By far the most common method of aligning amateur low noise receivers relies upon listening to a weak signal from a distant station, such as a beacon, and adjusting the matching components of the receiver input stage for maximum signal to noise ratio [1]. However, signals from distant sources are notoriously unreliable, varying rapidly in strength over a range of many dB. This makes it necessary to repeatedly check the strength of the beacon to ensure that an improvement in signal to noise ratio has been achieved.

A locally generated signal which can be adjusted in level down to barely detectable would appear to be ideal, since it would not suffer from the vagaries of propagation. However, in practice, it can be very difficult to attenuate the test signal down to the required level because of the amount of screening needed. Another issue with this approach is the matching between source and receiver: a well attenuated signal generator output will provide a good 50Ω match, whereas the antenna may not provide the same degree of matching and the resulting performance can be less than optimal.

A better approach to aligning low-noise re-

line to be fixed by one of the screws holding the coaxial socket. The general layout for this type of meter is shown in **Fig 8.26**.

→ For more designs, see also:

Test Equipment for the Radio Amateur, (3rd & 4th Edns), Clive Smith, G4FZH RSGB.

'Absorption Wavemeters for 144MHz', G R Jessop, G6JP, *Radio Communication*, Sept., 1982.

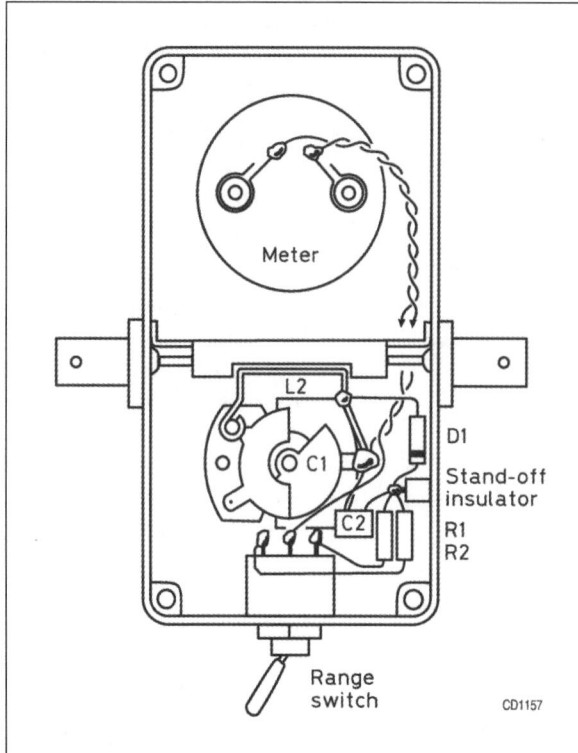

Fig 8.26 Arrangement of components for a directly coupled through-line wavemeter

ceivers is to use a noise generator in place of the signal generator [2]. With this technique, broadband noise is injected into the receiver input. The noise source is turned on and off, and the receiver matching adjusted until the ratio of 'noise on' to 'noise off' at the receiver output is at a maximum. It can be very difficult to judge aurally when the ratio is at a maximum, so some form of visual indication is desirable.

Such an instrument is known as an automatic noise figure meter when it indicates directly the true noise figure of the item under test. It is, however, necessary to use a source with an accurately known noise output in order to make an accurate measurement. If the noise output of the source is not accurately known the instrument can still be used to adjust the receiver for best signal to noise performance, although the actual noise figure cannot be determined. The instrument

described in the following sections can be used to adjust receivers operating at any frequency for optimum sensitivity. The instrument provides a continuous readout of the difference between the audio output of a receiver with no RF input, and the output when a wideband noise generator is connected to the receiver's antenna socket. The meter indicates the ratio between the outputs under these two conditions.

By design, the meter reading is not affected by changes in audio level over a wide range of volume settings. The circuit has a logarithmic response so that the meter scale can be linearly calibrated in signal to noise ratio in dB. Again, unless the absolute level of noise output from the noise source is known, the scale cannot be marked in noise figure. The unit uses a reverse-biased diode as a noise source.

Circuit description

The circuit diagram is shown in **Fig 8.27**, with the component listing in **Table 8.6**. Audio input from the loudspeaker socket of the receiver-under-test is connected to a small speaker (LS1) at the instrument input. This speaker provides a means of monitoring the receiver output which would otherwise be inaudible due to the muting action of most loudspeaker external connection sockets. The AC across the speaker is rectified by the precision rectifier formed around IC1. This arrangement effectively overcomes the forward-voltage diode-drop of a rectifier and thus very low level AC signals can be accommodated. D2 and R3 prevent the operational amplifier saturating on negative half cycles of the input. R4/C1 acts as a low-pass filter to the input of IC2.

IC2 is formed into a logarithmic amplifier by the use of TR1 in the feedback loop. Note: the voltage across the base/emitter junction of a transistor with its base connected to its collector is proportional to the logarithm of the current through the transistor.

Because the receiver is fed with two signals, IC1 and hence IC2 are also fed with these signals. The difference (in the mV range) between the output voltages under these two conditions is a function of the ratio between the two input voltages and this ratio is independent of the average input level. Provided that the various stages of the receiver and the circuit around IC1 are working within their linear range, the AC output from the circuit formed around IC2 at the pulse frequency used will be dependent only on the overall signal-to-noise ratio. As the output from IC2 is only a fraction of a volt peak-to-peak, it is amplified by the following stage formed around IC3.

The output of IC3 is fed to a unity gain phase-sensitive detector (PSD) based on IC4. The reference signal is fed via TR2 from the pulse generator (or multivibrator) TR3/TR4. A PSD is ideally suited to applications such as this, where an indication is required of the magnitude of an AC signal which has a known frequency and phase, but a high accompanying noise level. In this application, the PSD gives a usable output even when the signal is accompanied by so much noise that it is undetectable by ear.

IC4, which has relatively low output impedance, can drive a 1mA meter. Full scale deflection of the meter in the prototype was set at approximately 10dB signal-to-noise with the scale reading linearly in dB. R12/R13 limit the current through the meter with C3 providing smoothing of the detected signal, otherwise the meter would show an erratic response due to the nature of the noise inputs.

The pulse generator is formed from a conventional astable multivibrator (TR3/TR4) operating at about 30Hz, the output of this being fed to amplifier TR5 which is used to pulse the noise source.

The noise generator uses a reverse biased diode mounted in a separate enclosure with matching and decoupling components. An ideal arrangement would be to mount this within a coaxial plug. The diode, D3, used in the prototype was a CV364 microwave mixer, but alternatives are 1N21, 1N23, 1N25 and 1N32 or similar. A possible, but not tried, alternative is a BAT31 silicon avalanche device which is intended as a noise source from 10Hz to 18GHz.

Construction

Construction of the receiver alignment aid is not critical and audio techniques can be used, with the exception of the noise head which must be built using VHF techniques if it is to operate reliably at the highest frequencies. The circuit requires a symmetrical ±9V DC supply at about 20mA.

Alignment

The unit requires little alignment and no test equipment is needed. Plug the noise head into a receiver and gradually increase the diode current until an audible 'purring' sound is heard in the receiver loudspeaker. Connect the audio output of the receiver to the input of the unit. The 'purring' should now transfer to the unit's loudspeaker and the meter should show a fairly steady reading which can be varied by adjusting the noise-diode current (using RV3). Set RV1 so that the meter reading is constant over a wide range of receiver volume settings. Set RV2 to give a full scale deflection of the meter at maximum diode current on the highest frequency band of interest. The unit is now ready for use.

Operation

Connect the unit and noise head to the receiver under test and adjust RV3 for about half scale deflection on the meter. Any adjustment to the receiver that results in an improved signal gain with no change in the noise figure, or a reduced noise figure with no change in signal gain, or both simultaneously will result in an increased meter reading. By noting the reading of the meter before and after any circuit adjustments, improvements in performance can readily be seen.

Although the unit is not especially sensitive to small temperature changes, it is best to switch the unit on at least 10 minutes before use and to ensure that the ambient temperature is reasonably constant.

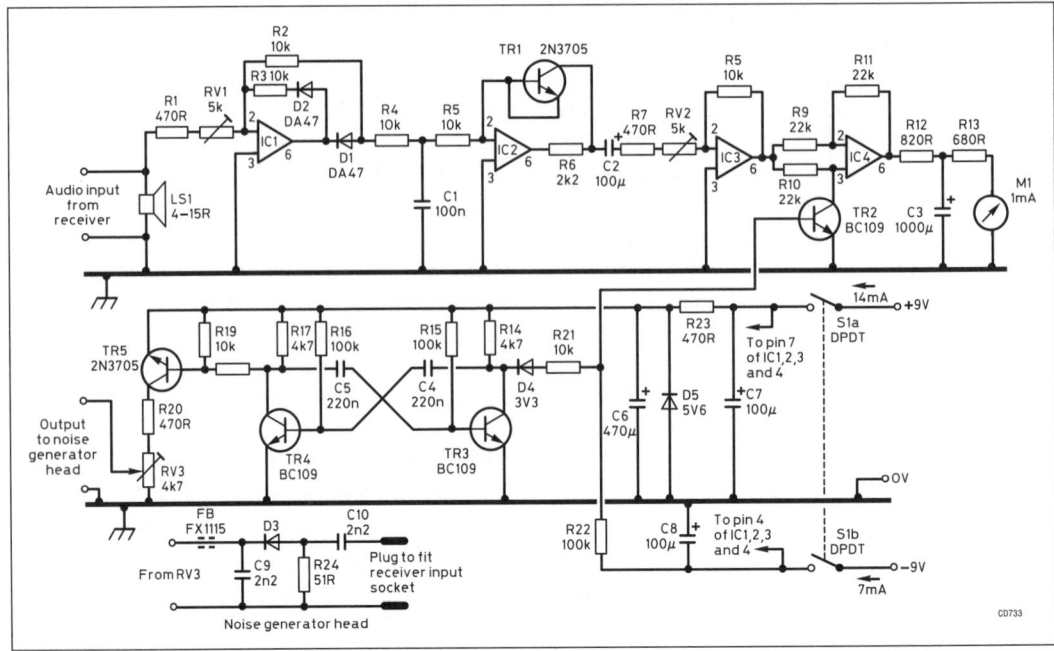

Fig 8.27 Circuit diagram of the alignment aid

Additional notes on use

Use of the receiver alignment aid assumes reasonable linearity of the receiver, therefore care must be taken when aligning FM receivers to ensure that the receiver does not limit with the noise source on. With most receivers this will mean that the level of noise injected must be as small as possible, consistent with still exceeding the FM threshold. With AM/SSB receivers the noise blanker and AGC must be disabled if meaningful results are to be obtained.

Care must be taken if the alignment aid is to be used for initial alignment of a converter or receiver. Noise output from the unit is constant over a wide range and it is therefore possible to align inadvertently on a spurious or image frequency, especially if the receiver has a low intermediate frequency. A signal generator or similar should therefore be used for initial alignment to avoid the problem.

Some receivers have been encountered that have a small DC voltage appearing at the loudspeaker socket. When connected to the alignment aid this voltage can bias IC1 beyond its linear range, thus resulting in false readings on the meter. Connecting an electrolytic capacitor of about 47µF in series with the input overcomes this problem. The negative terminal of this capacitor should be connected to the junction of R1 and the monitor loudspeaker.

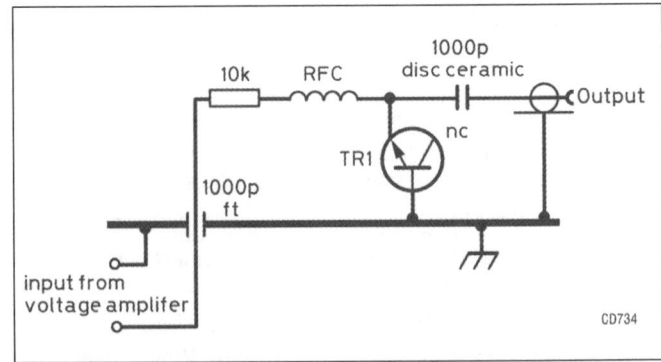

Fig 8.28 HF to UHF noise head. TR1 2N2369, BFY90 etc – see text. RFC is 3t of the 10k resistor lead, 2mm inside diameter

R1	470R	RV1,2	5k skel. preset, 0.1W	D3	see text
R2,3	10k	RV3	4k7 lin. carbon pot.	D4	3V3, 400mW zener diode
R4,5	10k	C1	100n, polyester	D5	5V6, 400mW zener diode
R6	2k2	C2	100µ, 6V3 tantalum		
R7	470R	C3	1000µ, 10V electrolytic	TR1,5	2N3705, 2N3703, 2N4126
R8	680k	C4,5	220n polyester	TR2,3,4	BC109, 2N2926
R9,10,11	22k	C6	470µ, 16V electrolytic	IC1-4	741 op-amp, 8pin
R12	820R	C7,8	100µ, 16V electrolytic	LS1	4 to 15R min. speaker
R13	680R	C9,10	2n2 ceramic disc	M1	1mA FSD meter
R14,17	4k7	FB	FX1115 or equivalent	S1	DPDT switch
R15,16	100k	D1,2	OA47, OA79, OA90, BAT85		
R18	22k				
R19,21	10k				
R20,23	470R				
R22	100k				
R24	51R or 75R				

Resistors are 0.25W/0.5W, 5% unless specified otherwise.

Table 8.6 Component list for the receiver alignment aid

Improvements

Considerable development work has been carried out to the receiver alignment aid since it was first published and this has resulted in several very useful improvements: see references [3] and [4].

The original noise head was designed primarily for VHF operation. An alternative design that can be used throughout the HF range and up to at least 1.3GHz is shown in **Fig 8.28**. Useful output may still be available at 2.3GHz when a suitable transistor is used for TR1. It is best to select a transistor with a high fT for TR1. It may be necessary to try several transistors before one with enough output is found.

Better phase detector performance is achieved at low levels with an FET (eg 2N3819) in place of the bipolar transistor TR2. Sometimes difficulties have been encountered with the meter reading not being independent of audio drive level. This can be just a matter of incorrect use or it can arise when the comparator is used in conjunction with receivers possessing an odd audio frequency response. This can be cured by replacing the components R4/C1/R5 by the circuit shown in **Fig 8.29**.

A slight improvement to the performance of the logarithmic amplifier may be obtained by

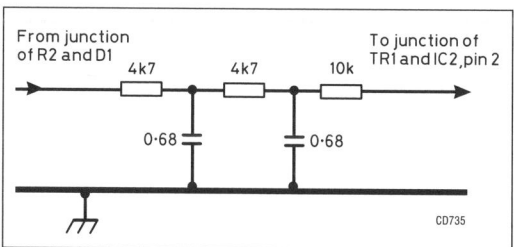

Fig 8.29 An improved inter-stage coupling network between IC1 and IC2

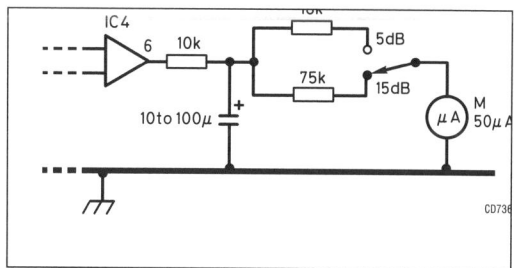

Fig 8.30 Modified meter circuitry

replacing R6 with a shorting link and increasing the value of C2 to 22 or 33µF. Fluctuating meter readings can also be a problem at times. Changing the meter to one of 50µA FSD and altering the time constant of the meter circuit can noticeably improve matters. **Fig 8.30** incorporates these modifications as well as including a switch to give different full-scale readings of signal-to-noise ratio.

[1] *Microwave Handbook*, Vol 2, Construction and Testing, edited by M W Dixon G3FR, RSGB, 1991.

[2] 'An Alignment Aid for VHF Receivers', J R Crompton, G4COM, *Radio Communication*, January 1976.

[3] 'Microwaves', C Suckling, G3WDG, *Radio Communication*, October 1979.

[4] '*Microwaves*', C Suckling, G3WDG, *Radio Communication*, March 1980.

→ See the following for noise sources to aid alignment:

- 'Noise Generator', *Solid State Design for the Radio Amateur*, pp167-168, ARRL, 1994.

- 'A Gated Noise Source' Chap. 25, *The ARRL Handbook for the Radio Amateur*, ARRL, 1988.

→ See the following for how to build a SINAD Meter:

'A Sinadmeter', Ed Chicken, G3BIK, pp87-91, *VHF Communication*, 2/2001.

'SINAD Meter', Ed Chicken, G3BIK, pp280-285, *VHF/UHF Handbook*, Andy Barter, G8ATD, RSGB.

Electrical Noise Detector

This project detects electrical radiation that causes problems to the amateur and allows the noise to be heard. When other members of the household are asked "please don't turn on that computer, vacuum cleaner or TV" and cannot understand why, this little device will allow them to see and hear the noise that amateurs have to contend with. The circuit (**Fig 8.31**, component list in **Table 8.7**) uses a

R1	1k	C10	100nF
R2,6	100R	IC1	741
R3,4	47k	IC2	LM386
R5	100k	LS1 Small 8-ohm speaker	
R7	10R		
RV1	10k, with switch	Miscellaneous:	
C1,6	4.7µF 16V electrolytic	3.5mm jack socket	
C2,5	10nF	Perforated board, 7cm x 7cm	
C3,4	22µF 16V electrolytic	Enclosure to suit	
C7	47nF	PP3 battery + clip	
C8	10µF, 16V electrolytic	Telephone pick-up coil	
C9,11	330µF 16V electrolytic	Resistors all 0.25W or similar	
Resistors are 0.25W/0.5W, 5% unless specified otherwise.			

Table 8.7 Component list for electrical noise detector

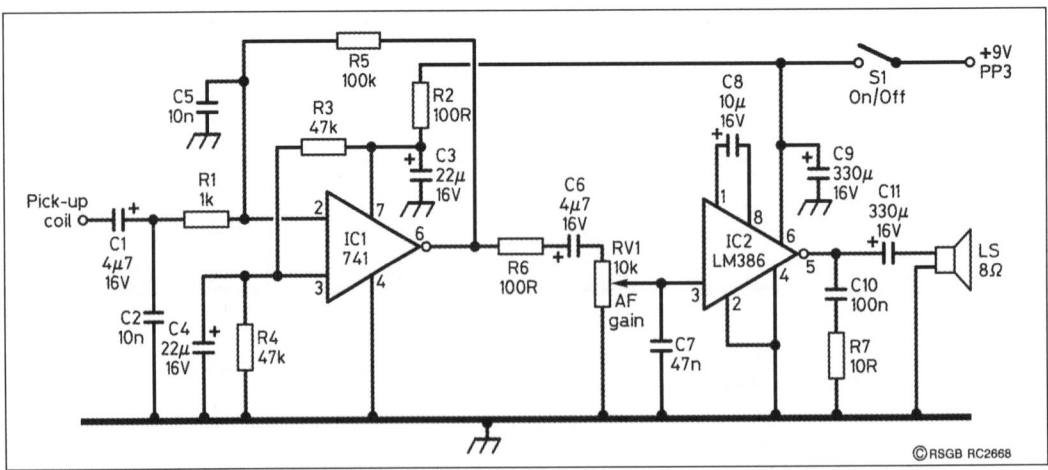

Fig 8.31 Circuit diagram of electrical noise detector

telephone pick-up coil as a detector, fed into a 741 IC pre-amp, followed by an LM386 power amp.

Construction

The project is built on stripboard with component leads pushed through the holes and joined with hook-up wire underneath; a wire running around the perimeter of the board forms an earth bus. Build from the loudspeaker backwards to VR1, apply power and touch the wiper of VR1. If everything is OK a loud buzz should be heard from the speaker. Too much gain may cause a feedback howl, in which case adjust RV1 to reduce the gain. Complete the rest of the wiring and test with a finger on the input, which should produce a click and a buzz. The pick-up coil comes with a lead and 3.5mm jack, so a suitable socket will be needed.

To demonstrate its utility, a high-impedance voltmeter set to a low AC voltage range was placed across the speaker leads to give a comparative readout between different items of equipment in the home. These readings obtained are shown in **Table 8.8**.

→ For the original article, see 'Electrical Noise Detector', Steve Ortmayer, G4RAW, p46, *RadCom*, July 2000.

→ For an item along similar lines, see 'Tx Monitor & Interference Sniffer', *Test Equipment for the Radio Amateur*, 4th Edn, Clive Smith, GM4FZH, RSGB.

29MHz oscilloscope	0.56V
Old computer monitor	0.86V
Old computer with plastic case	1.53V
New computer monitor	0.45V
New tower PC with metal case	0.15V
Old TV	1.2V
New TV	0.4V
Plastic-cased hair dryer	4.6V
Vacuum cleaner	3.6V
Drill	4.9V

Table 8.8 Readings obtained by placing the pick-up coil next to various household items

8.6 Loads, Attenuators & Matching Pads

Construction option: using plumber's fittings

There are various ways in which fittings used by plumbers, and cheaply available from local hardware stores, can be used to fabricate a dummy loads, attenuators and pads as well as samplers and couplers. An ideal item is the 15mm compression fitting, designed to join two sections of copper pipe of outside diameter 15mm. It has a slightly smaller section inside to stop the two pieces of pipe being joined from slipping through, and this is between 13.5 and 14mm diameter. This may need to be drilled- or bored-out for some applications. See **Pic 8.3** for some typical samples. The T-fitting is particularly useful for making samplers and couplers.

Connectors, such as N-type, BNC, or UHF, will need to be machined or cut so that they fit into the end of the compression fitting instead of the olive.

In the case of square-flange, chassis-mount connectors, turn (or file) the flange to fit just within the compression nut, so that it can be clamped between the nut and the compression fitting. Chamfer the small flange on the connector. Drill out the nut-hole fractionally to allow connector to protrude through.

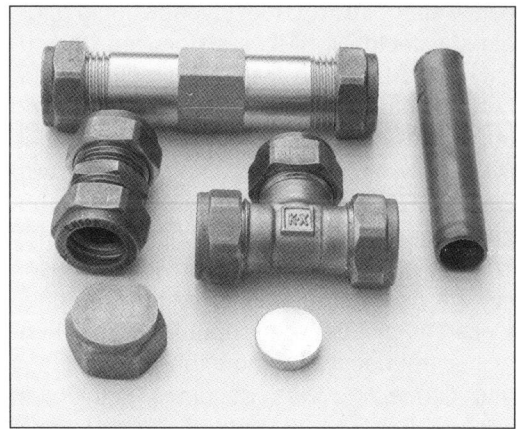

Pic 8.3 Typical plumbing items used

For round hole/nut-fixing, chassis-mount connectors, the N- and UHF-type will need some machining to fit inside the compression fitting and inside the nut; again chamfer the small flange on the connector. The hole in the nut needs making fractionally larger. The BNC type will need a disc making (or use a blanking disc) to fit inside the nut and a central hole to accommodate the connector; alternatively use a blanking nut. Sometimes it may be necessary to include the olive as well. Much depends on one's ingenuity and the extent of access to metalworking facilities.

Dummy Loads

An ideal dummy load absorbs all the power sent to it in a pure resistance, which is shielded to prevent radiation of the signal, and has a means of removing, or at least dissipating, the heat produced.

With home-brew, choice of components is crucial to performance. A key requirement is to use non-inductive resistors. Of leaded components, the carbon composition type is the best, but may only be available in 5% tolerance; these may well give useful performance into the low UHF region (eg 500MHz). Other types such as carbon film, metal film etc, rely on the resistive element being deposited on a former and then cut spirally to obtain the resistance value. This spiral-cutting may introduce some stray inductive effects at higher frequencies but at frequencies below about 30 MHz most types should operate satisfactorily. SMD resistors are particularly good because they are small and are not manufactured using any spiral-cutting techniques and hence the stray inductance is kept low. However, they can only cope with typical QRP operation. Higher power components based on thin and thick film technology housed in TO126, TO220 and TO247 cases can be used; their limiting factor is capacitance (1-2pF) between input leads, while at higher frequency, the inductance of the legs and connecting wire becomes significant. These allow construction of dummy loads for 144MHz, and with suitable constructional technique, up to the 70cm band.

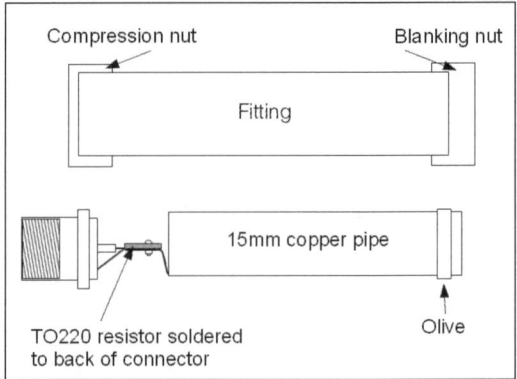

Fig 8.32 Construction of a dummy loads using plumber's fittings

Connectors chosen should be appropriate to the frequency range in use. Remember that the PL259/SO239 series is a non-constant impedance type and has degraded performance as frequency rises. Much amateur equipment from HF up to 144MHz has these connectors fitted as standard. Ideally, use good quality N, BNC or TNC type connectors, as these have good performance into the GHz range. Adapters can be used to convert from one connector type to another but these will all introduce reflections, however small.

As to fitting the load proper, an inner sleeve is made from 15mm copper pipe with a tab for the TO220 resistor (which is a 50Ω, 30W device). This tab is integral with the copper pipe

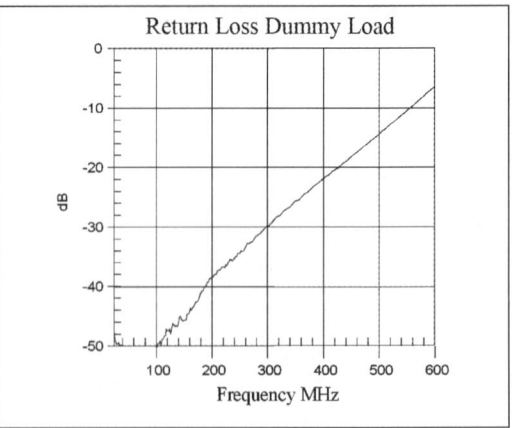

Fig 8.33 Frequency plot for a compression fitting dummy load

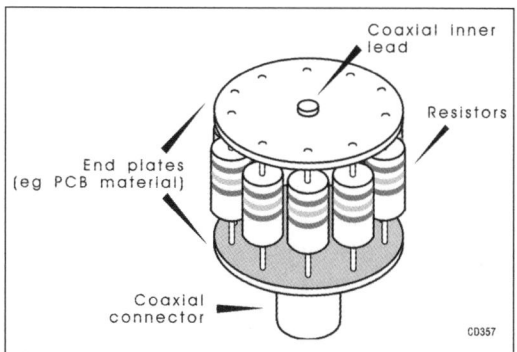

Fig 8.34 A multi-resistor dummy load

and is made by sawing the pipe vertically and longitudinally at one end and then bending the remainder to form the tab. The TO220 legs are bent slightly to allow the connector to be mounted centrally and mounted using a nut and bolt with thermal paste. Thermal paste is

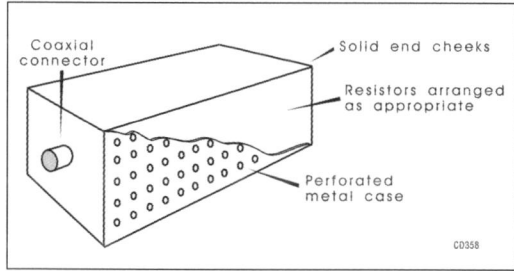

Fig 8.35 Possible construction of higher-power dummy load

also used on the outer of the copper pipe before it is slid into the compression fitting. At the non-connector end, the copper pipe protrudes so that an olive can just be fitted and then a blanking compression nut is used. Beware: when tightening up, do not let the inner pipe rotate with respect to the connector, as this will screw the resistor legs round.

It is not easy to replicate the heatsink typical of

a commercial product, which is often a finned case, perhaps with the resistive element inside strongly bound to it with thermally conductive paste, or immersed in oil. Clearly, some scheme for heat removal must be provided, if necessary by adding an external cooling, otherwise the load may overheat, changing its electrical properties and possibly destroying it. **Fig 8.32** shows a typical construction.

A frequency plot for the dummy load made in the compression fitting is given in **Fig 8.33**. The return loss is about 30dB (a VSWR of 1:1.07) at 300MHz and 18dB (1:1.29) at 450MHz. The unit is therefore adequate up to 70cm, and more than adequate at HF.

Multi-resistor dummy loads

To increase the power rating it is possible to use resistors in parallel; **Fig 8.34** shows a typical arrangement. These should, if at all possible, be encased in a metal shield to prevent unwanted radiation, or possibly a perforated shield to permit air flow - see **Fig 8.35**.

Arranging the resistors in a coaxial manner improves the perfomance of the dummy load. Ideally the pitch circle of the resistors and centre coaxial conductor should be carefully calculated, but as the size of this arrangement tends to be very short compared to a wavelength, non-adherence has little effect until the higher frequencies are reached. The characteristic impedance can be calculated using:

$$Z_0 = 138 \log_{10}(D/d)$$

where D is the pitch diameter of the resistors and d the diameter of the inner coaxial connector.

Typical arrangements for an approximately 50Ω load are given in **Table 8.9**.

The overall power rating is the sum of the power ratings of each resistor used. To obtain higher power ratings it may be possible to place the load resistors in a perforated screened container and air blow them to provide forced air cooling. This could be accomplished by placing a thermal switch on the resistors and using it

Resistor Value (Ω)	No in Parallel	Approximate Value (Ω)
100	2	50
150	3	50
390	8	49
560	11	51
1000	20	50

Table 8.9 Resistor values for 50Ω dummy loads

Pic 8.4 A higher power dummy load. (a) shows TO220 resistors on the heat sink - note use of heatsink compound for better heat transfer (b) shows the finished dummy load

to switch on a fan on once the temperature has risen above a certain point.

Sometimes large tubular carbon resistors come on to the surplus market, perhaps from firms such as Morganite, and these can make excellent dummy loads. Try to form them in a coaxial manner with the feed up the centre. Again, forced-air cooling can be used to increase power dissipation.

When using a dummy load, it may be possible to dissipate a much higher power for a short period of time, providing a long cool down period is allowed between applications of power. A commercial dummy load may often be provided with advice on this method of use.

Higher-power dummy load

TO220 package film resistors offer another method of achieving a dummy load. The main problem here is the inductance of the legs, which have an increasing effect as frequency rises, while the parasitic shunt capacitance also affects the input impedance.

However, with careful construction, including short device leads and an N-type socket to promote a low VSWR, operation to at least 100MHz should be possible.

Pic 8.4 shows the construction of such a dummy load. In this version, two 20W, 100Ω devices in parallel are used, although it is possible to use higher power resistors and different heatsinks. The heatsink shown fits within the base of a 89x28x35mm die-cast box, which provides RF shielding. The lid was used as the template for drilling the fixing holes.

Fig 8.36 shows the frequency response of this dummy load. Its return loss is better than 30dB below about 50MHz - a very satisfactory figure at HF.

→ Note a series of high power, high frequency (>1GHz) resistors by Bourns (types CHF-2010, 1206, 2525, 3725 and 3523) may be useful for making dummy loads.

Attenuators

A major problem when building an attenuator is associated with signal leakage, which can be between components of the attenuator, sections of the attenuator, or the input/output. It needs to be kept to a minimum and is caused by:

• Stray capacitance: This can occur between elements of the attenuator and can be significant in terms of performance, espe-

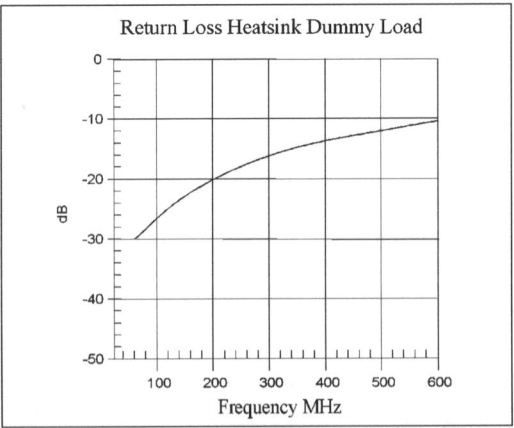

Fig 8.36 Frequency response of a higher power dummy load

cially when it occurs between the input and output of the attenuator. The effect is most noticeable at the higher frequencies.

• Stray inductance: Any lead can provide inductance as well as a path for inductive coupling. Like stray capacitance, this is particularly important in terms of coupling from input to output.

• Radiation: As the value of attenuation increases, the importance of a good enclosure becomes paramount. The attenuator housing should consist of a good RF-tight metal box, good shielding between stages, screening between the input and output, a good quality RF earth throughout and the use of high quality connectors. It is suggested that enclosures using die-cast boxes, plumber's fittings, PCB soldered boxes, and fabricated/machined items are best suited, as they give good screening. Do not use plastic type box constructions. See **Pic 8.5**.

With SMD technology in attenuator construction, excellent levels of performance can be achieved well beyond 1GHz. Thick-film types may give useful performance up to 8GHz, and thin-film, possibly to 12GHz. The problem with SMD technology is that of power dissipation – it tends to be good for low power receiver applications, but problematic where higher powers are encountered, as in transmitter use.

A key element of attenuator construction is not to have too high a level of attenuation in one section. High levels of attenuation in a single section may exacerbate the effects of stray inductance and capacitance, and any imperfections in earthing, causing the signal to bypass the attenuator. Hence, in attenuator construction, it is generally good practice not to attempt to achieve more than say 20dB of attenuation in a single attenuator section. When this is done, some of the resistors can be combined: in the case of the T-section the two inner series resistors can be added together, for the Pi-section the two inner parallel resistors.

Below are some suggestions for making attenuators and matching pads. Some of the dummy load techniques discussed earlier are applicable. The practical aspect requiring some ingenuity is the last solder joint before final assembly.

Calculation of resistor values

Attenuators and matching pads are normally made from Pi- or T-networks where there is a common earth throughout (unbalanced systems). For balanced systems there are H, O and Lattice networks. These variations are shown in **Fig 8.37**.

The choice of network may well be decided after resistor calculations are made, as the values of resistors are likely to be non-standard. Values will have to be made in practice from parallel and series combinations, but the fewer the number of resistors the better. It is always interesting, when resistor values have been chosen, to insert the values back into the equations to see the change in attenuation, and whether it is acceptable.

Pic 8.5 Various enclosures which give good RF-tight performance

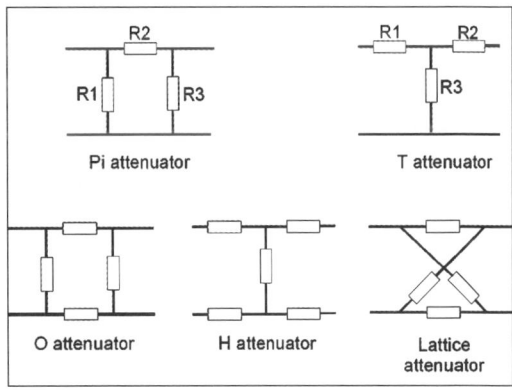

Fig 8.37 Various attenuator circuits

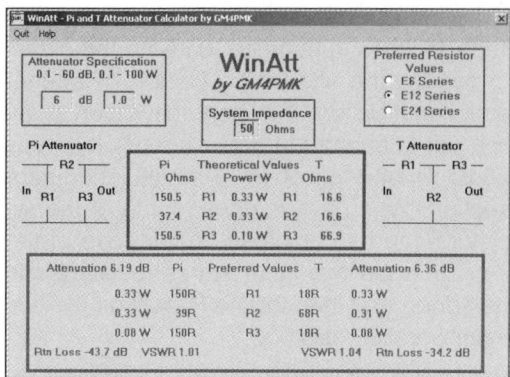

Pic 8.6 Screenshot of WinAtt by GM4PMK

The following formulae are for the unbalanced versions, as this is likely to be the choice used. If the required attenuation in dB is A, then the absolute attenuation a, is:

$a = 10^{A/20}$

For a Pi circuit:

$R1 = Z(a+1)/(a–1)$; $R2 = Z(a^2-1)/2a$

For a T circuit:

$R1 = Z(a-1)/(a+1)$; $R2 = 2Za/(a^2 -1)$

→ To make life easier, WinAtt.exe by GM4PMK can input all the requirements: attenuation, power rating, system impedance, Pi- or T-circuit, and resistor series to be used (E6, 12, 24), and output not only the exact value of the resistors required (and their power rating), but also the nearest preferred values, and the associated error in attenuation. A WinAtt screenshot is shown in **Pic 8.6**. To acquire this handy program, download the zip file at:

http://www.marsport.org.uk/winatt/winatt.zip

→ see similar calculators and lots of useful info at:

http://www.chemandy.com/calculators/matching-pi-attenuator-calculator.htm

http://www.giangrandi.ch/electronics/attenuators/attenuators.shtml

http://www.rfcafe.com/references/electrical/attenuators.htm

https://www.microwaves101.com/calculators/858-attenuator-calculator

https://www.random-science-tools.com/electronics/PI_attenuator.html

https://www.everythingrf.com/rf-calculators/attenuator-calculator

https://www.random-science-tools.com/electronics/T_attenuator.html

http://www.radio-electronics.com/info/rf-technology-design/attenuators/rf-attenuators-basics-tutorial.php

A simple 50Ω attenuator

This considers the use of Pi- and T-circuits for an unbalanced system. **Table 8.10** gives resistor values for 3, 6, 10 and 20dB attenuators plus suggested resistor combinations using the E24 series to give the required values. For values <10 these include first decimal place, otherwise they are rounded to the nearest whole number. Other combinations can of course be used.

It was decided to build a low power version of the 10dB attenuator in a 15mm compression

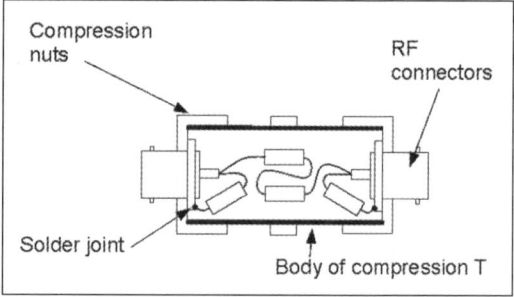

Fig 8.38 Diagram of attenuator construction

	T Attenuator		Pi Attenuator	
	R1, R2	**R3**	**R1, R3**	**R2**
3dB	8.5 (16+18 in parallel)	142 (120+22 in series)	292 (22+270 in series)	18
6dB	17 (2 x 33 in parallel)	67 (56 + 11 in series)	150	37 (33 + 3.3 in series)
10dB	26 (33+120 in parallel)	35 (13+22 in series)	96 (160+240 in parallel)	71 (15+56 in series)
20dB	41 (2 x 82 in parallel)	10	61 (10+51 in series)	248 (68+180 in series)

Table 8.10 Resistor values for T and Pi networks and various attenuations. It is preferable to use 1% resistors or to select them on test

Pic 8.7 Photograph of compression fitting attenuator

plumbing fitting, the ends being composed of a normal compression nut with blanking disc. The blanking disc was drilled to accept a nut-fixing BNC socket. The Pi version was chosen as it is easier using the two shunt resistors at either end and the series resistor to join it together. **Fig 8.38** is a diagram showing how the unit is constructed using leaded components, with 100R instead of 96R for the shunt resistors and two resistors (15R and 56R) for the series element.

The hard part is joining the two ends which can be fabricated externally. Make up the end connectors with shunt elements soldered to BNC pin and its base. Then make the two series resistors with a bit of overlap to allow a zig-zag in the series as shown in **Fig 8.38**. Solder this to one end and fit in the

compression body so that the end of the series resistor only just protrudes from the far end – cut if necessary and keep it very short. Solder the opposite BNC fitting to the series element and carefully push in, making sure it does not touch the compression body. A rolled-paper insulator could be used to help prevent a short. Tighten the end nuts, but do not allow the BNC connectors to rotate. If SMD resistors are used, solder one to each BNC pin and use a short piece of wire between them to facilitate the final construction. See **Pic 8.7** for the finished item. Don't forget to mark each unit with its attenuation factor. The attenuator has a return loss of 17dB (VSWR=1:1.32) at 435MHz, so it is certainly useful up to 70cm; the return loss drops to 10dB (VSWR=1:1.9) at 1GHz. The attenuation is fairly good at 10dB throughout.

A 50Ω Low Power Switched Attenuator

The circuit shown in **Fig 8.39** is a switched attenuator using a combination of Pi- and T-networks and enables preferred-value resistors to be used. A 1-2-4-8-...dB switching sequence is adopted so that the maximum attenuation range is obtained for a given number of sections. The switches are standard wafer-type panel-mount slide-switches, for

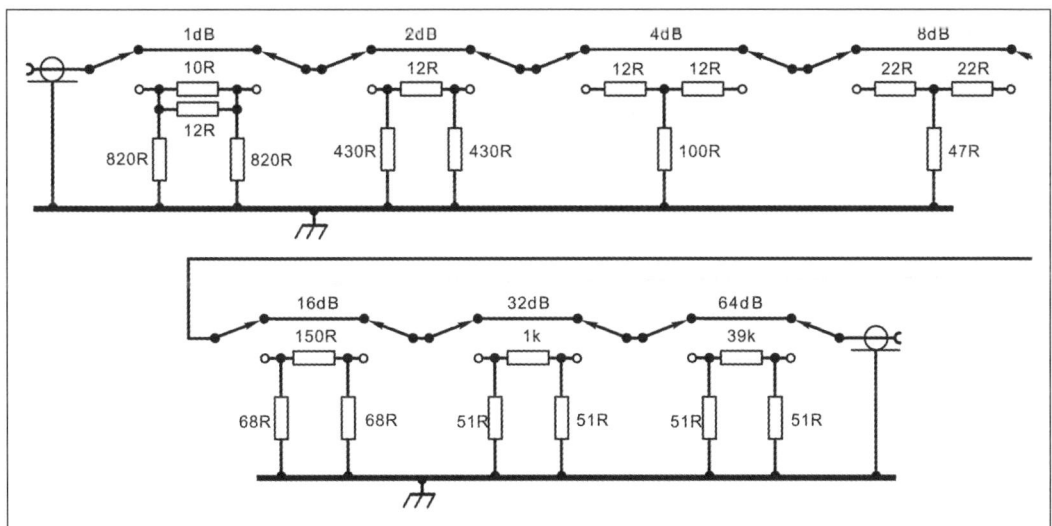

Fig 8.39 Low-power switched-attenuator using preferred resistor values

which the effective transfer capacitance in the circuit used is only 0.8pF. With 5% carbon film resistors the attenuator accuracy is ±0.5dB on the 1, 2, 4 and 8dB positions up to 500MHz. The 16dB position is 1dB low at 500MHz, the 32dB position 1dB low at 30MHz and the 64dB position 1dB low at 750kHz.

Matching Pads

There are various ways of fabricating a home-brew matching pad, but the essential requirement is for matching between unlike impedances and an RF-tight enclosure. Using resistive elements, matching is only achieved at the expense of attenuation. Practically, 6dB is the minimum attenuation one can get away with, 10dB may be a more useful value. Use an on-line calculator that allows for unequal terminating impedances to make the calculations, and then choose a combination of resistors to ease construction. Plumbing fittings can again be used and, as for the attenuator, the part requiring most thought is the last solder joint before boxing it all up.

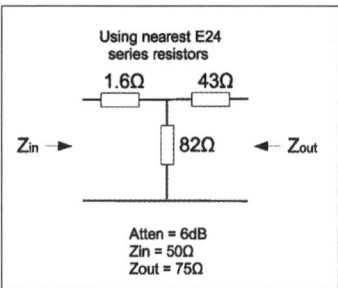

Fig 8.40 Circuit diagram of 6dB matching pad between a 50 and 75 ohm system

Consider a T-network that matches 50Ω to 75Ω with 6dB attenuation. The circuit diagram is shown in **Fig 8.40**, and **Pic 8.8** shows a possible method of construction. The body is made from hexagonal aluminium bar, bored

Pic 8.8 Photograph of matching pad

through and the ends drilled and tapped to accept fixing screws for the N-type connectors. The screw-plug on the body (top of photograph) enables the final internal connection to be made by being able to solder through the hole. For this version a machined part was made on a lathe at home.

Open- and Short-Circuit Loads

Occasionally, a short- or open-circuit load is required to calibrate equipment. These can be made quite easily using good quality N-type fittings and give good performance. **Pic 8.9** shows a photograph of the two units.

Short-circuit load

A small brass disc is made to fit inside the N-type on the top of the PTFE insulator with central pin. The disc is then drilled in the centre where an 8BA or similar screw will fit and go into the N-type pin. The disc is then

Pic 8.9 Photograph of N-type short- and open-circuit loads

inserted into the back of the N-type so that it sits on the ledge around the insulator and the screw is inserted right through into the pin. This is then soldered in place in the pin and to the brass disc.

So that the plug is fully enclosed, a 'plug' is made to be an interference fit into the N-screw collar and marked with an 'S' for short circuit. The sleeve (which holds the short in place) is then put in the back of the N-type and the collar screwed in. See **Fig 8.41**.

This short circuit looks good to greater than 1GHz with a return loss of less than 0.02dB.

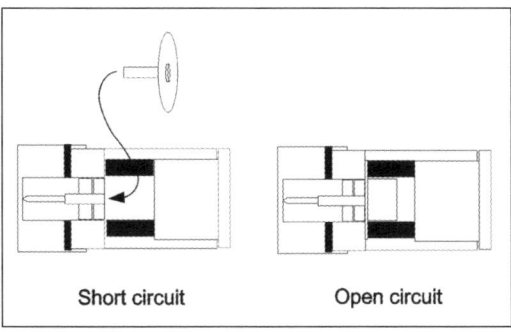

Short circuit   Open circuit

Fig 8.41 Cross-section of N-type short-circuit and open-circuit loads

Open-circuit load

This is simply an N-type plug with no cable. The coaxial bush is held in place by the sleeve and screwed collar. Again, see **Fig 8.41** for details. This load is good to greater than 1GHz.

8.7 Samplers and Couplers

There is a wide variety of possible practical implementations. The following are a few suggestions.

The T-Connector Sampler

As the name suggests, the generic arrangement looks like a 'T', because the input and output ports are in-line and sampled port comes off the centre. **Fig 8.42** shows the general arrangement and **Pic 8.10** a photograph of a practical construction using a small metal box. A sample of the signal going through is 'siphoned-off' using a simple inductive, capacitive or resistive coupling technique. The example shown here incorporates a resistive sampler (see below) and a Pi-circuit attenuator using E24 series resistors.

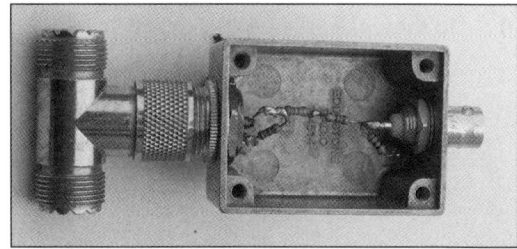

Pic 8.10 A practical T piece sampler

Tip: Try to use a different connector such as a BNC or SMA for the 'sampled' output to prevent inadvertent wiring/cabling and damage to the measurement/monitoring apparatus. Also remember that the PL259/SO239 series is a non-constant impedance type and not really suitable for use above about 150MHz.

Resistive sampling

Fig 8.43 shows a basic resistive sampler. As a rule of thumb, make the input resistance of the sampling network at least 10 times the characteristic impedance of the through-line to prevent unwanted loading. Use carbon resistors if possible, and determine their rating from the power on the through-line. Here, the power rating of the 9.1kΩ resistor is 0.5W for 100W through. This is CW; clearly lower ratings are possible for SSB operation as the mean power is less.

To protect test equipment, a series capacitor can be added to block any DC voltage, while back-to-back diodes can limit the input voltage, as in **Fig 8.44**. The rating is 5W continuous for this 1kΩ example.

Calculation of sampler attenuation requires knowledge of the input impedance of the

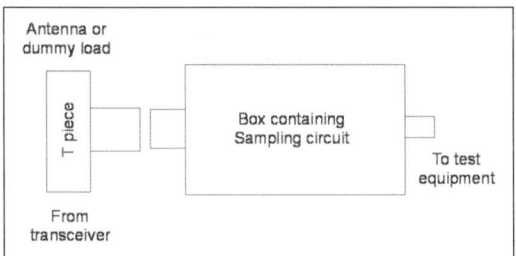

Fig 8.42 General arrangement of T piece sampler

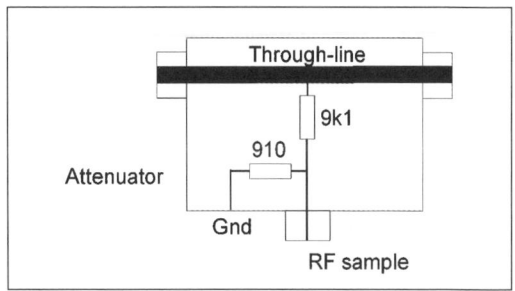

Fig 8.43 Resistive sampler

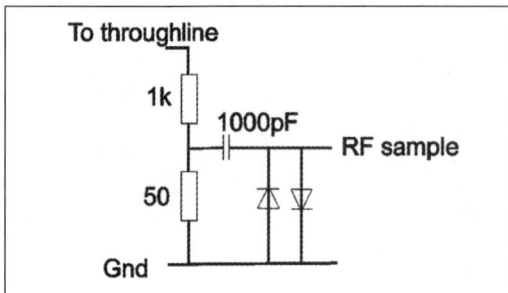

Fig 8.44 Typical circuits for resistive sampler

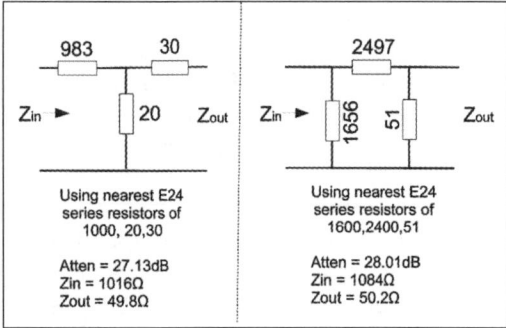

Fig 8.45 Resistive attenuator sampler circuits

equipment being attached to the sampled output and the impedance of the properly terminated through-line. Assuming that loading of the attenuator on the through-line is minimal (say 1kΩ) input and that the through-line is terminated in a 50Ω load, the resistor values for a Pi- or T-attenuator can be determined using an on-line calculator. Typical T- and Pi-circuits with theoretical component values are given in **Fig 8.45**. The input resistance of these networks is approximately 1kΩ and the output 50Ω. This is equivalent to a through attenuation of approximately 27dB. Power calculations must of course be made for the resistors used.

In calculating the attenuator circuits it is assumed that the test equipment wants a voltage drive, hence a figure of 1V on

the through-line provides 10mV at the test equipment terminals (ie a voltage reduction of 100:1).

If only approximate values of attenuation are required, use the nearest preferred values of resistances found in the junk box or combinations of series and parallel circuits.

Inductive Sampling

This method uses an inductive-coupling transformer arrangement for acquiring the RF sample. **Fig 8.46** shows (a) the theoretical circuit and (b) the practical arrangement. The turns-ratio and coupling factor determine the impedance seen by the through signal, which should be as high as possible. Again, only high input impedance test equipment, such as an oscilloscope probe should be connected to the sampled output. The sampler is typically constructed in a die-cast box 89x35x30mm (or equivalent) with end sockets linked by a length of 12SWG wire with a single turn loop at the centre. A two-turn coupling loop is taken to a BNC socket for connection to the test equipment.

Inductive coupling is the basis of many power and VSWR meters. These typically use the through-line as the primary of a transformer,

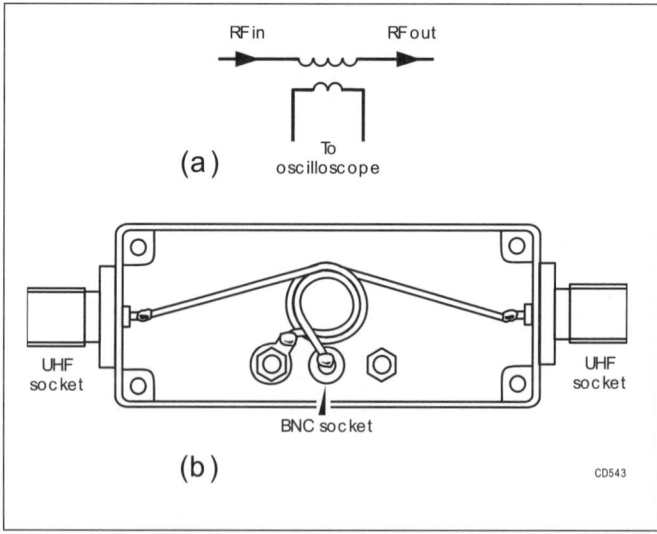

Fig 8.46 Basic inductive coupling for a sampler

and a loop (probably wound on a ferrite) as the secondary. The secondary provides the RF sample required. Note: no diodes etc for conversion to DC are required as this is an RF sampler.

The Through-Line Sampler

Here a probe is inserted into the field between the central conductor and outer of a coaxial system, with minimum disturbance to the line, and a signal is induced in the loop which can then be processed (**Fig 8.47**).

Fig 8.48 shows two examples of a coaxial system that can be easily constructed. The dimensions of the structures can be calculated using the following equations:

Theory

If a rod (diameter d) is inserted coaxially into a tube of diameter D, as in **Fig 8.48** (a), the characteristic impedance (Z_0) is given by:

$Z_0 = 138Log_{10}(D/d)$

If a similar rod is inserted into a square tube of equal sides D, as in **Fig 8.48** (b), the characteristic impedance is given by:

$Z_0 = 2 + 143Log_{10}(D/d)$

If now a sampling element is carefully inserted into the system without disturbing it too much, the Z_0 remains almost the same, but a sample of the signal can be obtained.

Homebrew Coaxial system with Internal Sampler

For the rectangular version, a die-cast box of internal cross-section 31x29mm (almost square) could be used. Taking the dimensions as a 28mm side, the internal conductor diameter needs to be 12.7mm for a 50Ω system and 8.5mm for a 75Ω system. For a 50Ω system, 12.7mm diameter is almost the diameter of a piece of 12mm copper pipe. So the through-line could be made from solid rod, or it could be fabricated using a section of 12mm copper pipe with a conical section at each end to make a smoother impedance transition to the connectors; the 8.5mm through-line could be made from 8mm copper pipe in a similar manner. The fact that it is hollow does not matter, as high frequency signals flows near the surface of the conductor, within the so-called 'skin depth'. The length must be cut to suit the die-cast box and the connectors used. **Fig 8.49** shows a typical implementation.

To make a sampling line, a small PCB line could be included where shown and a BNC socket attached. The PCB track is earthed at one end to the box using a nut and bolt. Countersink the screw-hole in the PCB so that the head does not intrude into the cavity. If necessary, cut the track at X and solder across with a 51Ω SMD resistor – this then provides a match to the test equipment. Alternatively, make a loop from say, 1.5mm (1/16in) brass rod and space it about 3.5mm from the wall of the die-cast box. Earth one end to the box, and the other end to the pin of the RF sample connector. The length is about the same as for the PCB track.

For the round version, it is convenient to use 15mm copper water pipe, whose inside diameter is about 14.5mm. From the above

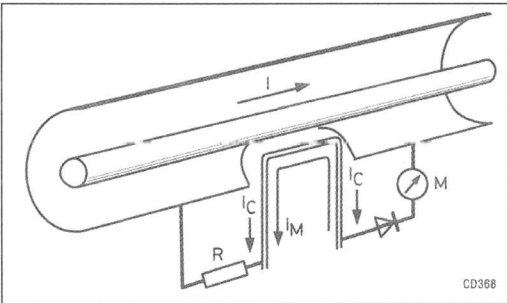

Fig 8.47 Coupling between coaxial inner conductor and sampling loop

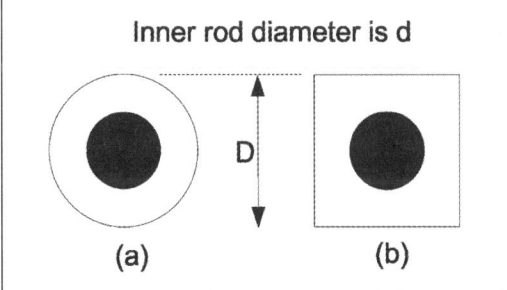

Fig 8.48 Two different arrangements for a coaxial system

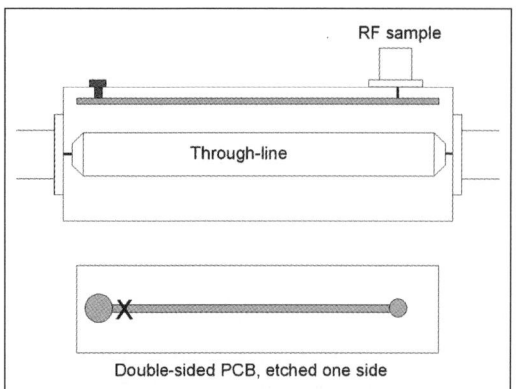

Fig 8.49 Typical construction of a coaxial sampler

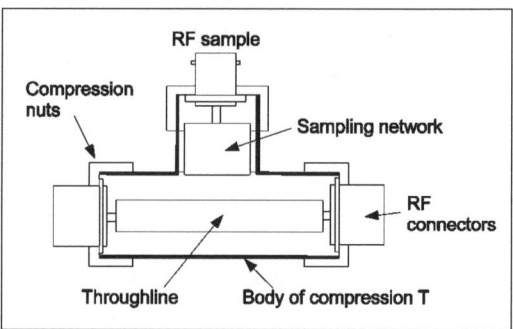

Fig 8.50 General arrangement of compression T-sampler

Pic 8.11 Compression T-sampler

formulae, this suggests an inner conductor diameter of 6.3mm for a 50Ω system and 4.1mm for a 75Ω system. A practical implementation could be to use a typical 15mm compression T-joint obtained from a plumbing supplier, and a 6.3mm (say 6mm or ¼in) rod some 33-35mm long. If there happens to be a stop in the T fitting, it might be possible to drill the centre of the T to 14mm. See **Fig 8.50** for the general arrangement and **Pic 8.11** for a photograph of a typical unit.

The connectors will need to be machined or adapted so that they fit into the end of the compression fitting. First decide on the connector type (eg N, BNC or UHF) for the through-line and then the connector for the test equipment connection. For safety reasons the test equipment connector should be different to the through-line connectors.

The next item to be given some thought, is how to connect the through-line itself to the back of the connectors and the way to provide a good electrical connection. The following is for N-type connectors: adapt the ideas for other connectors.

The end of the centre conductor needs drilling to a depth of 3mm in order to take the pin on the back of the N-type connectors and should allow a tight push-fit for the pins. Assemble the connectors in the compression T and see if any needs filing from the end of the centre conductor. When of the correct length (or fractionally greater) the main through-line

can finally be fitted and the compression nuts tightened to hold the through-line in place. All that needs doing now is to make the sampling element. There are two ways of doing this: by using *capacitive coupling*, or using *inductive coupling*. In both cases the probe should be positioned to give about a 2mm clearance from the centre conductor, which will require accurate measuring of the 15mm T that is being used.

Capacitive coupling: This requires a short stub to be fitted to the back of the sampling connector with a disc at the end to fit symmetrically into the compression fitting. This can be fabricated from a brass screw. Use the head to provide the disc and cut the shank to the appropriate length. Drill the end to accept the pin of the 'sample' connector and solder on. Check that there is at least a 2mm clearance between disc and through-line. **Fig 8.51** shows an example, with the head of the screw in line with the internal body of the compression fitting.

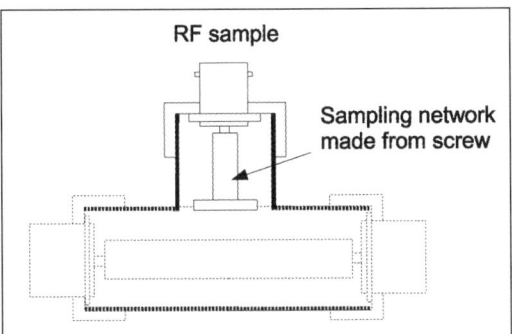

Fig 8.51 Capacitive 'sampling' head

Inductive coupling: Here one needs to sample the magnetic field contained within the sampler. Again care must be taken not to touch the centre conductor. **Pic 8.12** shows a typical construction before assembly; a piece of stripboard (single sided) is cut to a disc 15mm diameter – just enough to fit snuggly on to the shoulder in the top of the T. Cut the disc so that there is a hole at the centre to line up with connector pin. Trim the board so that the copper tracks do not touch the brass of the T fitting, then make a loop using wire, which for safety can be sleeved with silicone sleeving. Allow this to protrude through the matrix board as far as the connector and then solder in place. A 50Ω resistor (use 4/Ω or 51Ω) can then be soldered to the central strip, and the central conductor also soldered: this provides a 50Ω source impedance for test equipment connection. The sampling loop can then be soldered to the connector, the ground connection being to

the body and the centre to the pin. Insert the finished sampler into the plumber's T with the matrix strips at right angles to power flow in the main part. Make the unit so that the bayonet lugs of the BNC fitting are in line with the sampling loop as this makes installation much easier.

Simple sampler using a coaxial cable

Thread a length of insulated wire between braid and polythene, or through an air spaced coaxial cable - each has only a small effect on the coaxial transmission line. The ends without braiding should be kept as short as possible and good connectors used. The ends of the sampling line can then be used appropriately and the unit housed in a metal box. See **Pic 8.13** for a sampling line using UR67/RG213 cable.

A BNC socket has been provided for the output signal. If a directional unit is required then obviously another socket can be provided together with another sampling line.

Other possibilities include:

• Semi-airspaced cable: sampling wire fed through an air cavity

• Heliax: here a small hole could be melted longitudinally through the dielectric (hot knitting needle?) near the outer shield, such that a small wire could be pushed through to create the 'sampling' loop. Alternatively a slot could be ground in the corrugated outer and a small wire probe inserted.

Pic 8.12 Pre-assembly photograph of inductive sampling head

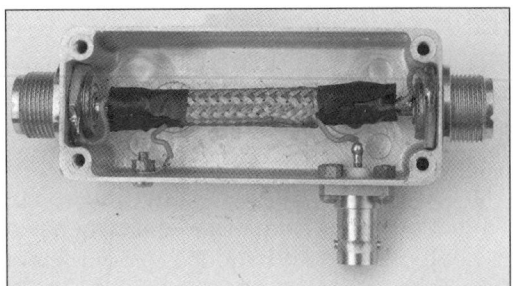

Pic 8.13 Basic arrangement of coaxial sampling line

8.8 Directional Couplers

A number of practical schemes for making directional couplers are provided, some with detection facilities for power measurement.

A coaxial directional coupler

This coupler has a single coupling port and must be reversed to measure reflected power. The measured characteristics are given after the constructional details. The coupler is made using a round conductor in an almost square section die-cast box (**Fig 8.52**). The box used was 89x35x30mm and the

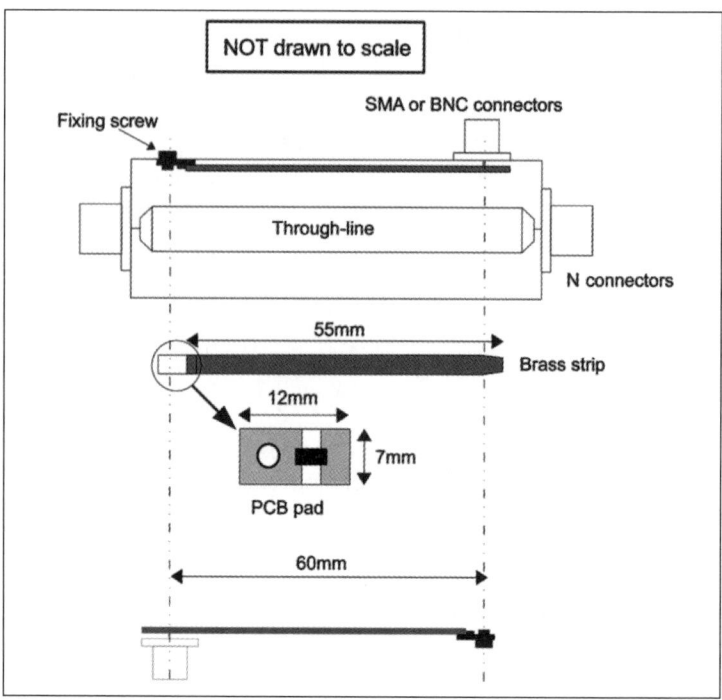

Fig 8.52 Directional coupler construction

through-line made from brass rod, calculated at 13mm diameter, although 12.5mm (½in) is probably adequate; as would be a brass tube with a suitable end fitting. The through-line was 77mm long and chamfered at each end over the last 5mm. The ends were drilled centrally 3mm diameter to provide a push fit for the connector pins, but these can be soldered after construction is complete.

Because of the corner fixing posts in the die-cast box, the N-type flange screws were made self-tapping as it is awkward to use a nut and bolt. If obtainable, N-type sockets with round-hole and nut-fittings are easier to use. N-type connectors for the through path are fitted centrally to each end of the die-cast box, whereas the coupled output connector is SMA. The fixing screws are 60mm apart fitted centrally on the box side.

The coupling loop was made from brass strip, 7mm wide and 0.3mm thick - this dimension is not too important but don't exceed about 0.5mm. Several widths of strip were tried, but with the spacing off the die-cast box side of the FR4 PCB of thickness 1.6mm, the 7mm wide strip gave best results. This approximates to a 50Ω stripline).

The FR4 PCB pad was 1.6mm thick and a cut made on the copper of 2mm wide to isolate two lands. The brass strip was then soldered to the narrower end of the pad and a 51Ω SMD resistor (shown in black) soldered across the cut as shown. The pad was drilled to accommodate the fixing screw (typically 2mm or 8BA). **Pic 8.14** shows a photograph of a finished unit with the lid removed.

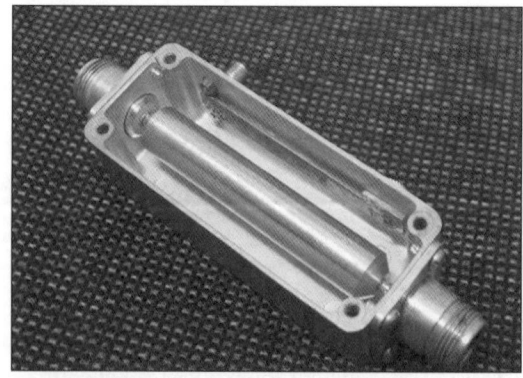

Pic 8.14 Inside the directional coupler

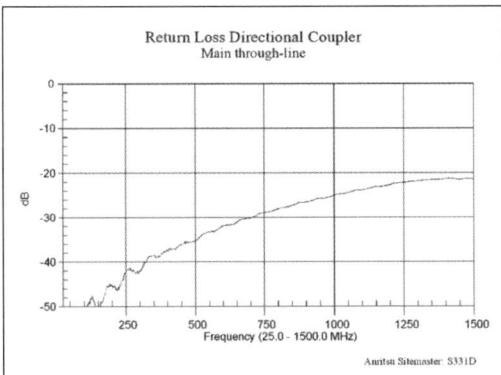

Fig 8.53 Return loss of directional coupler

Performance: Two sets of measurements were made; the first was the return loss of the main through-line (**Fig 8.53**). The second were of forward and reverse coupling factors and the difference, the directivity (**Fig 8.54**).

The return loss up to 1500MHz is better than 22dB (VSWR=1.2:1) while 435MHz and below is better than 37dB (VSWR=1.03:1). When the line was reversed similar results were obtained. The directivity is always better than 20dB and almost 30dB at 50MHz. The forward coupling factor is 20dB at 1GHz and 40dB at 50MHz giving power ratios of 1/100 and 1/10000 respectively.

Further development: If an identical coupling loop (shown in grey in **Fig 8.52**) is inserted on the opposite side of the main through-line then a directional coupler is formed with both forward and reflected outputs available simultaneously. The unit can be converted to a coupler/detector by the addition of external components such as diodes, capacitors and resistors - see next item.

→ For: A General-purpose directional coupler/detector, and a PCB coupler/detector

see Chap. 11, *VHF/UHF Handbook* (2nd Edn), Andy Barter, G8AT-D,RSGB 2008.

→ See also: 'High Power Directional Couplers with Excellent Performance', Paul Wade, W1GHZ, *Dubus* Vol39, 2/2010.

8.9 Reflectometers and VSWR Meters

Some RF samplers and couplers have resistors, diodes, and capacitors attached in order to detect the sampled RF signal and convert it to DC. This DC voltage is a measure of the power being sampled, and if both forward and reverse are measured, the VSWR can be obtained.

A Reflectometer for VHF and UHF

This unit will indicate effective forward and reflected power from which the VSWR can be obtained. However, the meter deflections are roughly proportional to the frequency so that it is only useful for power measurement if calibrated against another power meter for each frequency band. The instrument as dscribed is intended for use at 2m and 70cm as the meter deflection may be inconveniently small at lower frequencies.

Construction: To couple to the inner conductor, a section of the transmission line must be enlarged so that the directional couplers can be inserted. The enlarged section may be of either round or square cross section but the latter is easier to construct unless access to a lathe is possible. The impedance of the new section of the line must, of course, be the same as that of the normal feeder line.

Fig 8.55 shows the general arrangement. In this, the bottom is closed by a suitable plate and the ends by plates fixed to all four sides of the square section. In order to preserve

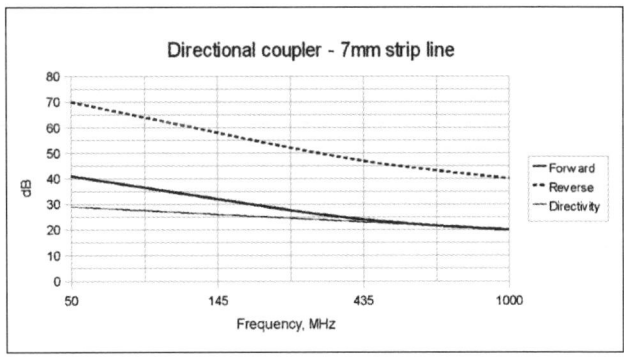

Fig 8.54 Forward and reflected coupling and directivity of the directional coupler

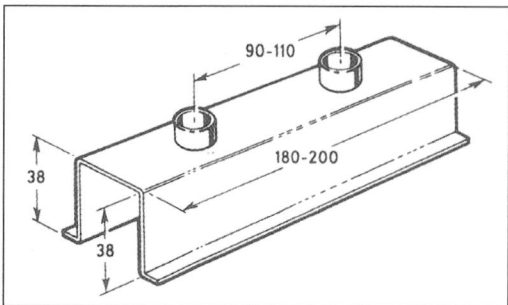

Fig 8.55 Square section of line showing detector head locations. Dimensions in mm

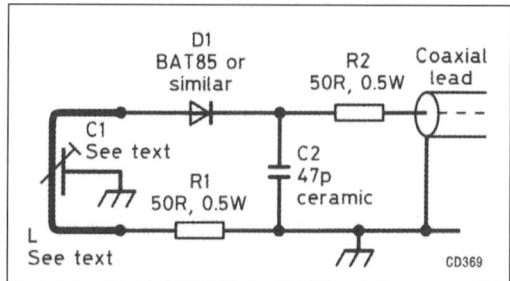

Fig 8.57 Circuit of detector head

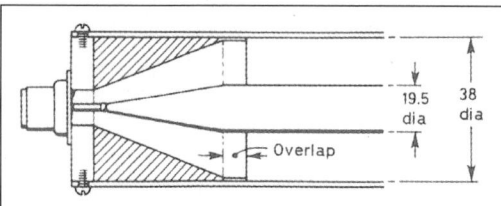

Fig 8.56 End tapers of both inner and outer of the section of line. All dimensions in mm

the impedance at each end of the enlarged line, tapers of constant D/d should be provided as shown in **Fig 8.56**.

Note: the inner conductor may be shaped by turning or fitting a thin copper cone rolled and soldered to the inner conductor; the outer taper can be made by either fitting blocks in the ends or folding thin sheets into the correct shape. If this latter method is used, overlap should be allowed for adequate contact to the outer.

Detector head mounting: the detector heads and sockets are made from telescopic tubes of about 24 and 25mm diameter, the sockets being 6.5mm long and spaced as shown in **Fig 8.55**. The sockets are soldered into holes cut to fit them, and are slotted so that they may be clamped tightly to the detector heads. The clamps can be made from 6.5mm wide brass strip bent round and held together with a 6BA nut and bolt.

Detector heads: The circuit for each detector head is given in **Fig 8.57** and the general constructional details on **Fig 8.58**. The signal received on the loop (L) is rectified by D1, smoothed by C2 and passed to a meter via the coaxial lead.

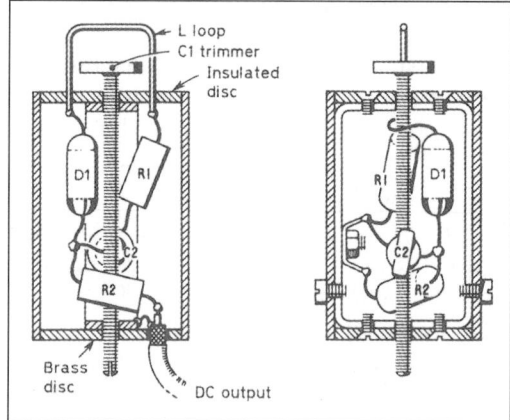

Fig 8.58 Two cross-sectional views of detector head

The heads are constructed on small rectangular frames made of 6.5mm wide thin brass strip, bent into rectangles which fit snugly into the tubes. It is upon this chassis that all the components are mounted. The upper end of each head is closed by a brass disc and the bottom end is made from a disc of insulating material. These end plates are held in position by 8BA screws which are also used for fixing the chassis inside the tube. It is important that the heads should be a good fit in their respective tubes so that the whole assembly can be easily pushed together.

The small trimmer capacitor C1 is made by screwing a piece of 6BA studding right through the rectangular end plates. A 9-10mm diameter disc is then soldered to one end, the other slotted to take a screwdriver. The assembly is used to adjust the

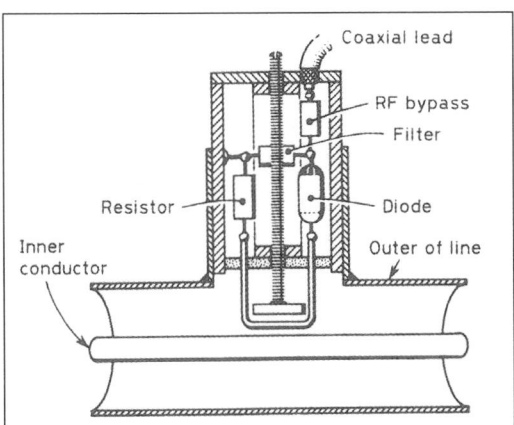

Fig 8.59 Mounting of detector head

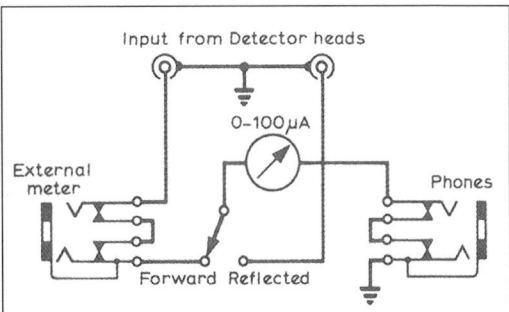

Fig 8.60 Circuit diagram of the meter switching for a mounted reflectometer

capacitance between the loop and ground. The coupling loop is made from 1mm (approx 20SWG) wire, and is 12.5mm wide and about 10mm above the insulated disc on which it is mounted. It is fixed through pairs of small holes 12.5mm apart, each end of the loop being passed through one hole and folded back through its neighbour and then pinched to make a firm anchorage for each leg. The mounting of the head in the line is diagrammatically shown in **Fig 8.59**.

Terminating resistance: the initial setting up process requires that a terminating resistance is used at one end. It must look like a resistance, even at 432MHz, which means that is must be coaxial and of similar value to the line. This can be either a commercial unit of suitable frequency range or can be constructed. The resistor should have the same DC value as the required terminating resistor and be of a carbon type (about 1W). It is used as the inner conductor of a coaxial line. The diameter of the outer conductor should be somewhat smaller than that given by the usual formula for the characteristic impedance of coaxial lines in order to compensate for the reactive nature of the inner - a ratio of diameters of 1.5:1 would be suitable for a 50Ω line.

Setting-up: a low power oscillator/transmitteramplitude modulated with a continuous note should be fed into the reflectometer

with the terminating resistor at the other end. *Note*: use only low power because the terminating resistor may only be 1W. After this setting-up procedure the unit can of course be used at full power.

The detector heads should be inserted so that the loops are about half way between the inner and outer conductors of the line, using the clamps as depth controls. Plug a set of headphones into the circuit of **Fig 8.60** to connect to the detector heads, and a signal will be heard; select the head nearest the terminating resistor.

The detector head should now be rotated in its mounting and the trimmer adjusted until a sharp null is heard in the headphones. Now clamp the detector head. The input and output connections to the line should now be reversed and the headphones switched to the opposite head and the same procedure followed. This procedure can be carried out on 144MHz but may need slight adjustment for 432MHz.

The unit is now ready for use. If required, it can be calibrated against a commercial power meter so that the micro-ammeter scale can be read directly in watts.

A Broadband VSWR meter

Reflectometers designed as VSWR indicators have normally used sampling loops capacitively coupled to a length of transmission line - this results in a meter deflection that is roughly proportional to

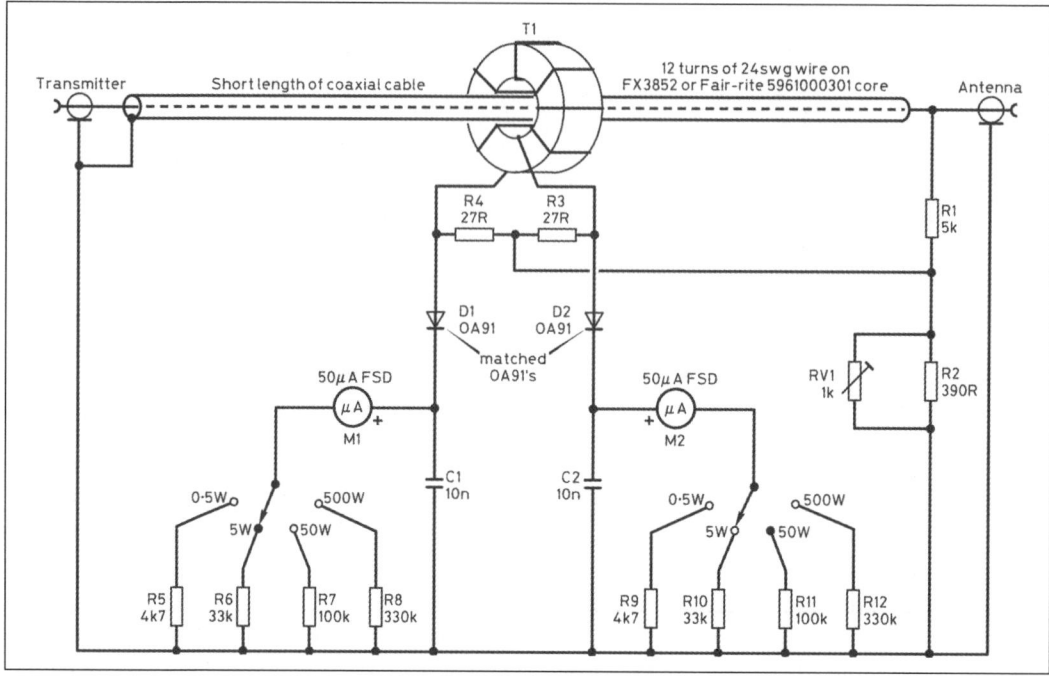

Fig 8.61 Circuit of the frequency-independent VSWR meter

frequency and they are therefore unsuitable for power measurement unless calibrated for use over a narrow band. By using lumped components, this shortcoming can be largely eliminated and the following design may be regarded as independent of frequency up to about 70MHz.

This instrument is for a 50Ω line and has full-scale deflections of 0.5, 5, 50 and 500W selected by two range switches. These should not normally be ganged since the reverse power will normally be much less than the forward power. The outer of the coaxial cable acts as an electrostatic screen between the centre conductor and secondary winding of the transformer; the cable length is not important over the frequencies it is intended to operate over.

Circuit description

The circuit is shown in **Fig 8.61** and uses a current transformer in which the low resistance at the secondary is split into two equal parts, R3 and R4. The centre section is taken to the voltage-sampling network (R1, R2,

RV1) so that the sum and difference voltages are available at the ends of the transformer secondary winding.

Layout of the sampling circuit is fairly critical. The input and output sockets should be a few inches apart and connected together with a short length of coaxial cable. The coaxial cable outer must be earthed at one end only so that it acts as an electrostatic screen between the primary and secondary of the toroidal transformer. The layout of the sensing circuits in a similar instrument is shown in **Pic 8.15**.

The primary of the toroidal transformer is formed by threading a ferrite ring onto the coaxial cable. Twelve turns of 24SWG (0.56mm) enamelled copper wire are equally spaced around the entire circumference of the ring to form the secondary winding. The ferrite material should maintain a high permeability over the frequency range to be used: the original used a Mullard FX1596 which is no longer available but suggested alternatives are Philips FX3852 or 432202097180 and Fair-rite 5961000301. Other types may also be suitable.

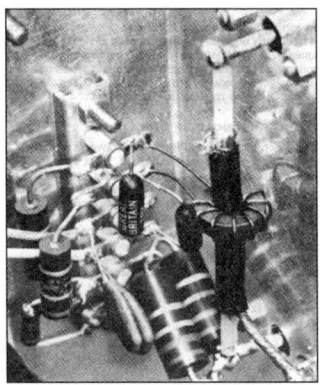

Pic 8.15 View of the sensing elements of the frequency-independent VSWR meter

The remaining components in the sampling circuits should have the shortest possible leads. R1 and R2 should be non-inductive carbon types. For powers above about 100W, R1 can consist of several 2W carbon resistors in parallel. RV1 should be a miniature skeleton potentiometer in order to keep stray reactance to a minimum. The detector diodes D1 and D2 should be matched point-contact germanium types with a peak inverse voltage rating of about 50V - OA91 diodes are suitable. The resistors R3 and R4 should be matched to 5% or better.

The ratio of the sampling resistors R1 and R2 is determined by the sensitivity of the current sensing circuit. As the two sampling voltages must be equal in magnitude under matched conditions, RV1 provides a fine adjustment of the ratio.

Germanium diodes as specified are essential if an instrument is to be used at low power levels, otherwise silicon diodes such as 1N914, or Schottky diodes such as the BAT85 may be substituted. To increase sensitivity at low power levels, eg 1W, the feed line could be looped through the toroid, in which case it may be necessary to use a larger toroid or a smaller coaxial cable (but this will not cope with high powers!) A components list for this project is given in **Table 8.11**.

Calibration

Accurate calibration requires a transmitter and an RF voltmeter or possibly an oscilloscope. The wattmeter is calibrated by feeding power through the meter into a dummy load of 50Ω. RV1 is adjusted for minimum reflected power indication and the power scale

R1	5k carbon (see text)
R2	390R carbon
R3,4	27R, 2W carbon
R5,9	4k7
R6,10	33k
R7,11	100k
R8,12	330k
RV1	1k skeleton pot 0.5W
C1,2	10n ceramic
T1	Philips FX3852, 4332202097180 Fairite 5961000301 with 12t, 24swg ECW
D1,2	OA91 matched (see text)
M1,2	50µA FSD meters
S1,2	1 pole, 4w rotary switch

Table 8.11 Component list for VSWR meter

calibrated according to the RF voltage appearing across the load. The reflected power meter is calibrated by reversing the connections to the coaxial line.

→ For a modern take on measuring VSWR using a resistive Wheatstone bridge and diode detection, but with the VFO replaced by a direct digital synthesis (DDS) module driving an Arduino microcontroller with LCD, see: 'Arduino-based SWR Analyser', Michael Booth, G8HKS, *RadCom*, April 2017.

8.10 RF Bridges

Noise Bridge for Measuring Antenna R and X

The circuit described here is an adaptation of that described in the second edition of this book. It allows a modulated signal to be obtained, if desired, by pulsing the supply to the noise generator [1]. Having the modulation ON may aid detection of the balance point, especially if an AM receiver is used. The circuit consists of a wideband noise generator followed by a bridge for making the measurements. The bridge allows the measurement of the parallel components of an unknown impedance to be measured. The circuit requires 9V DC at about 25mA.

[1] *The ARRL Handbook for the Radio Amateur*, post 1988, Chapter 25.

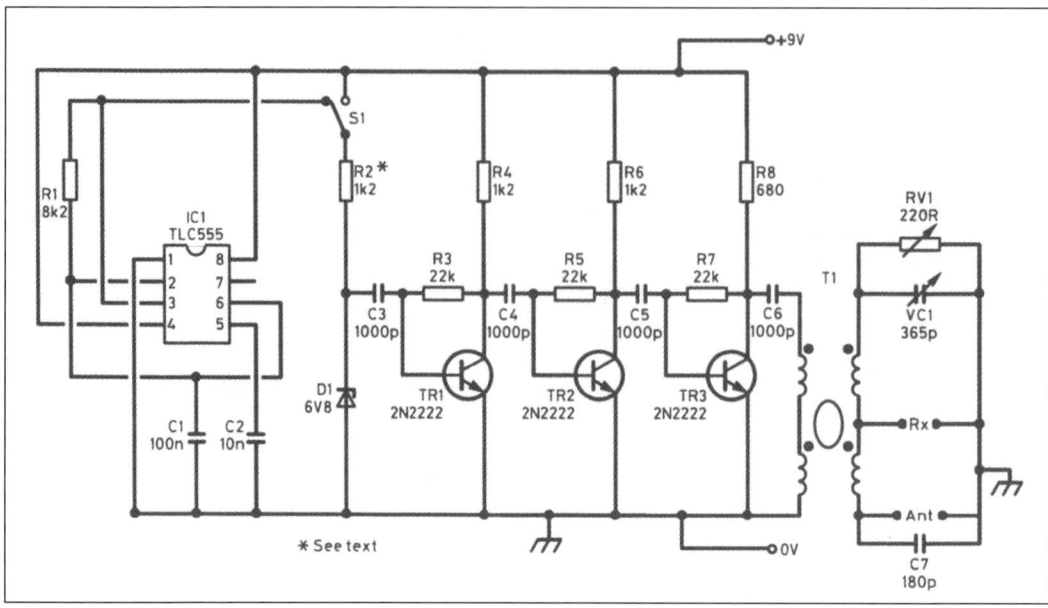

Fig 8.62 Circuit diagram of the modulated RF noise bridge

Circuit Description

The circuit is shown in **Fig 8.62** and the component listing in **Table 8.12**. The white noise is generated by the Zener diode D1 operating at low current. It may be possible to maximise the noise by suitable choice of the Zener diode and R2. The frequency range of this noise should extend up to at least 200MHz. The noise source is followed by a three stage wideband amplifier to raise the noise level to the order of 100μV. This enables a receiver to be used as a null indicator.

The noise output from the amplifier is applied to a quadrifilar wound toroid which forms the transformer T1. This provides two arms of a bridge circuit which has a variable resistor and capacitor in the third arm to obtain a balance against the antenna in the fourth arm. The bridge circuit is shown diagrammatically in **Fig 8.63**.

When the noise across the RV1/VC1 arm equals the noise across the antenna/capacitor combination, the bridge is said to be balanced, this occurs when the received noise signal is at a minimum. The values can be obtained from the settings of RV1 and VC1. The inclusion of C7 allows an offset to be used so

that inductive reactance can be measured. The midpoint setting of VC1 is equal to zero reactance. If a noise bridge is only required to measure the resistive part of the antenna impedance omit C7 and VC1. Timer IC1 is in astable mode and runs at about 850Hz with 50% duty cycle, this can be used to provide current for the Zener circuit via SW1 and thus modulates the noise source. The Zener

R1	8k2	
R2*	1k2	
R3,5,7		22k
R4,6	1k2	
R8	680	
RV1	220R pot, carbon	
C1	100n, 50V, ceramic	
C2	10n, 50V, ceramic	
C3,4,5,6	1000p, 50V, ceramic	
C7	180p, silver mica	
VC1	365p, Jackson, type 01 Gang	
TR1,2,3	2N2222	
T1	FT50-6, Dust iron core, 4 windings, each 14T; or 596100001 Ferrite core, 4 windings, each 6T	
D1	6V8 Zener, 400mW	
IC1	TLC555	
S1	SPCO switch	
Resistors are 0.25W/0.5W, 5% unless specified otherwise.		

Table 8.12 Component list for RF noise bridge

diode can be alternatively fed from the constant voltage power supply line.

Construction

The toroid transformer consists of a dust-iron core, type T50-6, which is wound as follows. Cut four lengths of 26SWG enamelled copper wire about 120mm long, twist them together and then thread them through the toroid to give 5 or 6 turns evenly spaced to cover the circumference. Divide the turns into two pairs, each pair consisting of two windings connected in series, the end of one winding connecting to the start of the other - be careful. Check that the two pairs are insulated from each other. Endeavour to keep the lead lengths in the bridge as short as possible and symmetrical. The variable resistor RV1 should be of high quality and with a carbon track - not wirewound!

When constructing the circuit, ensure that the noise generator and amplifiers are well away or screened from the bridge transformer and measuring circuit. The potentiometer case should not be earthed and, if it has a metal spindle, this should be isolated from the user and not contact ground. The complete circuit should be mounted in a screened box such as a die-cast type with appropriate connectors such as UHF-type or BNC. In order to avoid coupling into the measuring circuit of noise by way of current in earth loops, the earthed side of the noise source should not be joined to the general chassis earth of the bridge but should be taken by an insulated lead to the frame of the variable capacitor. As in all high frequency measuring circuits, lead inductances should be kept to an absolute minimum and where any lead length more than a few millimetres is unavoidable, copper foil at least 6mm wide should be used. All earth returns should be taken to the capacitor frame. Capacitor C7, which should be silver mica, can be soldered directly across the 'unknown' socket. A suitable PCB pattern and component layout is given in Appendix D.

Calibration

Connect a test resistor (non-inductive type) across the unknown socket with the receiver tuned to 3.5MHz. Adjust RV1 and VC1 to give a null. The value of RV1 is at the position equal to the test resistor, and the capacitor should be at approximately the mid-mesh position (the zero reactance condition): mark these positions. Repeat with different values of test resistor up to 220Ω in order to provide a calibration scale for RV1. Repeat this operation with known values of capacitance in parallel with the test resistor, up to a maximum value of 180pF. Mark the corresponding null positions on the VC1 scale with the value of this capacitance. Repeat this procedure at 28MHz to check the accuracy of the bridge. If the layout has been carefully attended to there should be little difference in the null positions.

To calibrate VC1 for negative capacitance values (ie inductance) it is necessary to place given values of capacitance temporarily in parallel with VC1. Gradually decrease the value of these capacitors (C_T) from 150pF towards zero, obtaining null positions and marking the VC1 scale with the value of $-(180-C_T)$ pF, ie if 100pF is substituted, then the negative C value is 80pF.

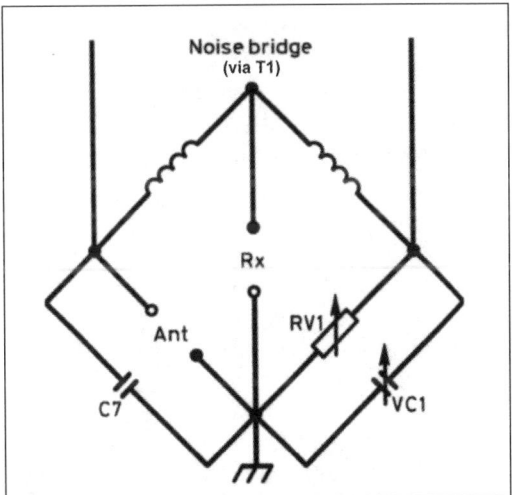

Fig 8.63 Diagrammatic representation of noise bridge

Using the noise bridge

For work on an antenna, a noise bridge should ideally be connected across the antenna terminals. This is usually not practical, in which case a noise bridge should be connected to the antenna by a length of line which is a multiple of a half wavelength at the frequency of interest (taking into account the velocity factor of the cable).

Connect the impedance to be measured to the 'unknown' socket, switch on the noise generator and tune the receiver to the frequency at which the test is to be made. Use RV1 and VC1 to obtain a minimum noise reading on the receiver S-meter. The values must now be converted to circuit components. The value recorded from RV1 is the resistive part of the impedance. The value from VC1 is the parallel reactive component of the impedance and depending on the sign is either inductive or capacitive. If it is positive, the value of shunt capacitance can be read directly from the VC1 scale. If it is negative, the VC1 reading represents the value of the shunt inductance can be calculated using the formula:

$L(\mu H) = 1/(4\pi^2 f^2(180-C))$ where f is in MHz and C in pF

→ See also 'Noise Bridge Measurements' by Brian Horfall, G3GKG, *RadCom*, April 2003

RF Impedance Bridge

The instrument described here will measure impedances from 0 to 400Ω at frequencies up to 30MHz. It does not measure reactance or indicate if the impedance is capacitive or inductive. A good indication of the reactance present can be obtained from the fact that any reactance will mean a higher minimum meter reading.

Circuit description

There are many possible circuits, some using potentiometers as the variable arm and others variable capacitors, but a typical cir-

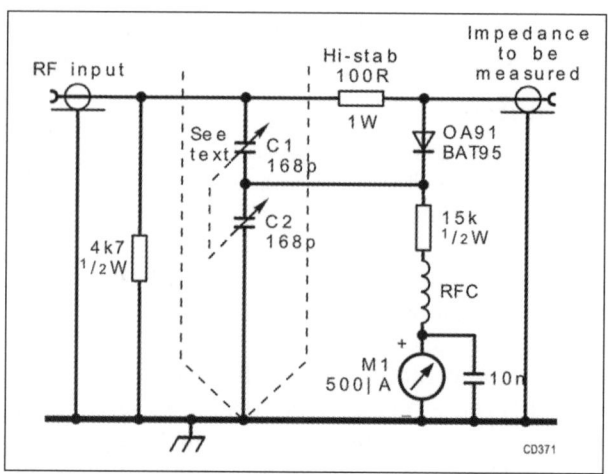

Fig 8.64 Simple RF bridge. Note that a BAT85 diode may be used instead of the OA91

cuit is shown in **Fig 8.64**. The capacitors have to be differential in action, and mounted in such a way that as the capacitance of one decreases the capacitance of the other increases. The capacitors should be of the type which have a spindle protruding at either end so that they can be connected together by a shaft coupler. To avoid hand-capacitance effects, the control knob on the outside of the instrument should be connected to the nearest capacitor by a short length of plastic coupling rod. These capacitors form two arms of the bridge, the third arm being the 100Ω non-inductive resistor and the fourth arm, the impedance to be measured. Balance of the bridge is indicated by a minimum reading on the meter M1. Construction is straightforward, but keep all leads as short as possible. The unit should be built into a metal box and screening provided as shown by the dotted line in **Fig 8.64**.

Signal source and calibration

The instrument can be calibrated by placing across the load terminals various non-reactive resistors (ie not wire-wound) of known value. This calibration should ideally be performed at a low frequency where stray capacitance effects are at a minimum, but calibration holds good throughout the frequency range. In using the instrument, it should be remembered that an exact null will only be obtained on the meter when the instrument has a purely resistive load. When reactance is present, it becomes obvious from the behaviour of the meter, as adjusting the control knob will give a minimum reading, but a complete null cannot be obtained.

The RF input to drive the bridge can be obtained from a dip oscillator, signal generator, or low-power transmitter capable of giving up to about 1W of signal power. The signal source can be coupled to the bridge by a short length of coaxial cable directly, or via a link coil of about four turns as shown in **Fig 8.65**.

If using a dip oscillator, care should be exercised in order to not over-couple, as it may pull the frequency or, in the worst case, cause it to stop oscillating. As coupling is increased it will be seen that the meter reading of the bridge increases up to a certain point, after which further increase in coupling causes the meter reading to fall. A little less coupling than that which gives the maximum bridge meter reading is the best to use. The bridge can be used to find antenna impedance as well as for many other purposes, such as to find the input impedance of a receiver on a particular frequency.

Some practical uses

One useful application of this type of simple bridge is to find the frequency at which a length of transmission line is a quarter- or half-wavelength long electrically (see again **Fig 8.65**). If it is desired to find the frequency at which the transmission line is a quarter-wavelength, the line is connected to the bridge and the far end of it is left open-circuit. The bridge control is set to zero ohms. The dip oscillator is then adjusted until the lowest frequency is found at which the bridge shows a sharp null. This is the frequency at which the piece of transmission line is one quarter-wavelength. Odd multiples of this frequency can be checked in the same manner. In a similar way the frequency at which a piece of transmission line is a half-wavelength can also be found but in this case the

remote end should be a short-circuit. The bridge can also be used to check the characteristic impedance of a transmission line. This is often a worthwhile exercise, since appearances can be misleading. The procedure is as follows:

1. Find the frequency at which the length of transmission line under test is a quarter-wavelength long. Once this has been found, leave the oscillator set to this frequency.

2. Select a carbon resistor of approximately the same value as the probable characteristic impedance of the transmission line. Replace the transmission line by this resistor and measure the value of this resistor at the pre-set frequency. (Note: this will not necessarily be identical with its DC value).

3. Disconnect the resistor and reconnect the transmission line. Connect the resistor across the remote end of the transmission line.

4. Measure the impedance now presented by the transmission line at the preset frequency. The characteristic impedance (Z_0) is then given by:

$$Z_0 = \sqrt{(Z_s Z_r)}$$

where Z_s is the impedance presented by the line plus load and Z_r is the resistor value.

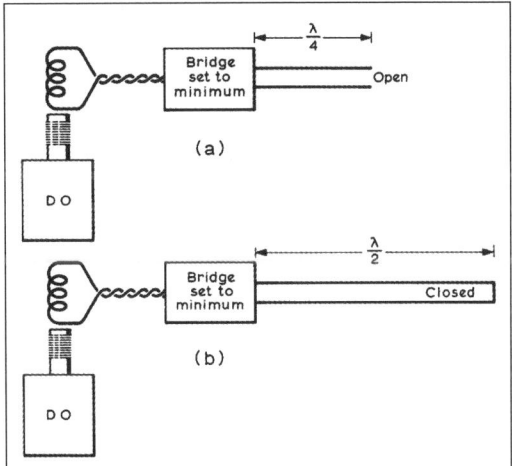

Fig 8.65 Use of the RF bridge with a dip oscillator

A Sensitive Antenna Bridge

This bridge (**Pic 8.16**) enables antenna systems to be adjusted for minimum VSWR using single-frequency 'CW' signal powers as low as a few microwatts over a frequency range of 1.8MHz to greater than 60MHz, and should be useable on 70MHz. It is most useful during extended periods of experimentation, where very low power transmissions should be used to avoid inconvenience to other band users and possible exposure of the experimenter to high (RF) voltages or powers. This instrument is not intended to supplant the in-line VSWR meter used for monitoring during normal transmissions. Its function is essentially the same as a 'resistance only' noise bridge, but it works with a

Pic 8.16 The frequency-independent VSWR meter

low-power external signal source. It uses a built-in moving-coil meter to monitor the signal and does not require a receiver. The bridge is so simple that it can be built in a few hours. Suitable signal sources include a signal generator, crystal oscillator, dip meter, and transmitter plus attenuator.

An outline specification is as follows:

• Operating signal power: -25 to +3dBm (3µW to 2mW) at 10MHz for meter FSD

• Frequency response: Less than 2dB variation from 1.8 to 65MHz

• Power supply: 9V PP3 battery with current drain of 18mA.

→ For full article and constructional information, see 'A Sensitive Antenna Bridge', Ian Braithwaite, G4COL, *RadCom*, July 1997, or the description in the previous edition of this book: *Test Equipment for the Radio Amateur*, (4th Edn), Clive Smith, G4FZH RSGB.

→ Elsewhere on this subject, see some useful practical notes in 'Homebrew', by Eamon Skelton, EI9GQ on: 'A Simple QRP SWR Bridge', 'A Simple Return Loss Bridge', and 'RLB Construction', *RadCom*, April 2017.

→ More generally see: 'Practical tips for building a return loss bridge', *VHF Communications*, 3/2009, Andrea Daretti, IZ2OUK.

8.11 Field Strength and RF Current Meters

Simple Field Strength Meter

A simple example is shown in **Fig 8.66**. A signal is picked up by the antenna, rectified and smoothed by D1/C1, the resulting DC signal is then indicated on the meter M1, with RV1 acting as a sensitivity control. Capacitor C2 provides an AC short across the meter for any unwanted RF signals.

Construct the unit in a box, with an external telescopic whip or a loop of wire. The unit can be used for relative field strength measurements at a given frequency, but not for relative measurements between different frequencies as the efficiency of the antenna and rectifier will affect readings. It should be

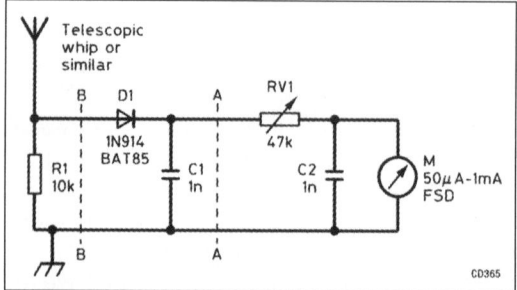

Fig 8.66 Simple un-tuned field strength meter

a useful device for tuning a transmitter to obtain maximum radiated power or adjusting an antenna for maximum radiated power. By splitting the circuit at AA, the antenna/rectifier combination could be used as a remote reading head, with the meter/sensitivity control being in the shack.

Higher Sensitivity Broadband Field Strength Meter

The concept here is to amplify the received signal, detect it, and drive a meter. This can be accomplished by using a relatively inexpensive broadband RF amplifier such as the MAR series from Mini Circuits placed in the circuit of **Fig 8.66** at position BB. Constructed on a circuit board with good ground plane, all components should have short leads to minimise lead inductance, while the capacitors should be carefully chosen for the frequency range envisaged. The active devices general-

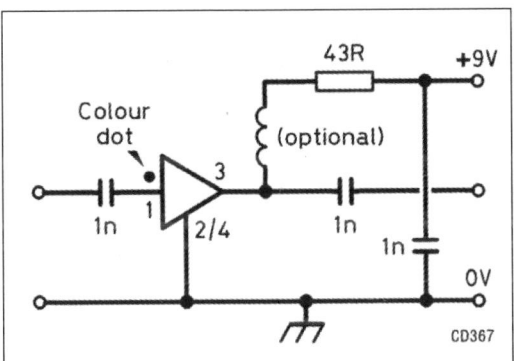

Fig 8.67 Broad-band amplifier based around a MAR8 or similar MMIC

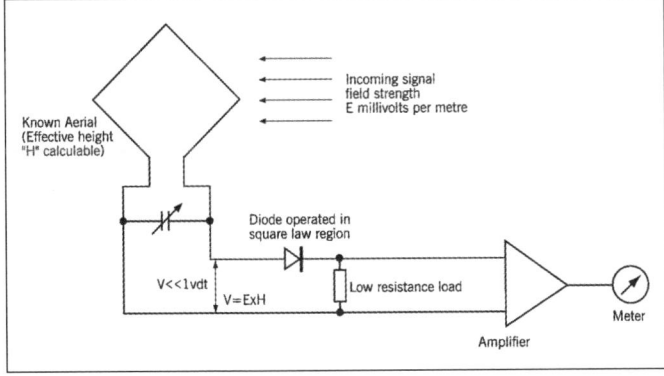

Fig 8.68 Concept of the field strength meter

ly have outputs of the order of +10dBm, so it is imperative that diode D1 is of a type with a low forward volt-drop, such as a Schottky type BAT85. If the field strengths being measured are very low it may be possible to cascade two or more such amplifiers. **Fig 8.67** shows a circuit based around a MAR8 which has a response from DC to 1GHz with a quoted gain at 100MHz of 33dB and 23dB at 1GHz; the maximum output is about +10dBm. The device requires 7.5V at 36mA, the circuit shows a series resistor for operation from a 9V DC supply. Other MMIC amplifiers can be used.

→ See also *The Amateur Experimenter's Guide*, 2nd Edn, Peter Dodd, G3LDO, RSGB 1996, where other circuits are suggested, as well as the use of a communications receiver.

A Calibrated Field Strength Meter

This equipment (**Fig 8.68**) uses a small 'known antenna', the characteristics of which can be easily established. This is followed by a very sensitive diode-detector and meter amplifier.

→ For full article see 'A Calibrated Field Strength Meter', David Sumner, G3VPH, *RadCom*, May 2006, or the abridged version in the previous edition of this book: *Test Equipment for the Radio Amateur*, (4th Edn), Clive Smith, G4FZH RSGB.

An RF Current Meter

The basic version of this handy device takes about 10 minutes to tack-solder together (**Fig 8.69**). When you're convinced how useful it is, you can then go on to build a more permanent version. The clip-on RF current meter has a long history; early versions involved breaking a ferrite ring into two equal pieces - which takes some doing! The constructional breakthrough was the idea of David Lauder, G0SNO - to use a large split ferrite bead intended for HF interference suppression. It clamps around the conductor under test, to form the one-turn primary of a wideband current

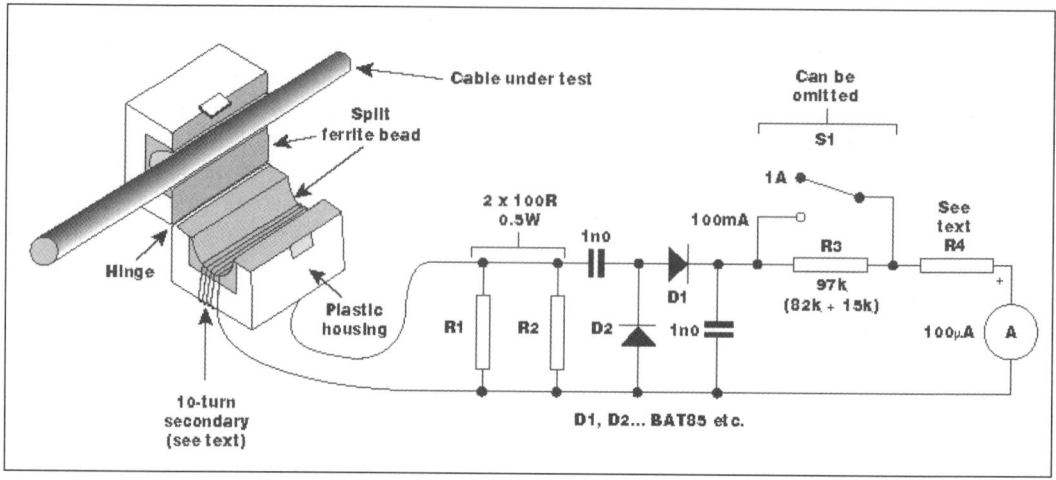

Fig 8.69 G0SNO's clip-on RF meter

transformer. The secondary winding is about 10 turns, and is connected to a load resistor, R1-R2, and the diode detector.

The load resistor, R1-R2, is important because it creates a low series impedance when the current transformer is effectively inserted into the conductor under test. For the values shown in **Fig 8.69** (10-turn secondary, 2 x 100Ω) this is 50/102 = 0.5Ω. Some circuits omit this resistor, but that creates a high insertion impedance - exactly the opposite of what is needed. Also, more secondary turns create a lower insertion impedance, but at the expense of HF bandwidth.

The other components in **Fig 8.69** are discussed in G0SNO's article which is reproduced on the 'In Practice' website (see link below).

Component types and values are critical only if you want to make a fully calibrated meter with switchable current ranges. However, for a first try, and for most general RFI investigations, the meter is almost as useful without any need for calibration. Make R4 about 4.7-10kΩ, and omit R3 and S1. If the meter is either too sensitive, or not sensitive enough, either change R4 or change the HF power level.

Just about any split ferrite core intended for RFI suppression will do the job, but there are

a few practical points. Choose a large core, typically with a 13mm diameter hole. This allows you to clip the core onto large coax, mains, or multi-core cables while still leaving enough space for the secondary winding (which should be made using very thin enamelled or other insulated wire). It is important that the core closes with no air gap - and that can be a problem.

A major disadvantage of the basic split ferrite core in its plastic housing is that the housing is not meant to be repeatedly opened and closed, so the hinge will soon break. By all means try out this gadget in the basic form but it is likely that you will soon be thinking about something more permanent. The classic way to do this is using a clothes peg (**Pic 8.17**), but there are now better alternatives.

For example, the first photograph in **Pic 8.17** shows the rather heavy-duty version using two strong clothes-pegs, fibreglass sheet and epoxy glue (more details at [1]). The second photograph shows Alan Chin's, GI0XAC, neat and simple version using a giant plastic paper-clip, with a small plastic-cased meter stuck on the side. The only requirement about the clip is that it must be non-metallic, and be able to hold the two halves of the core accurately together while the whole weight of the meter is dangling from the cable.

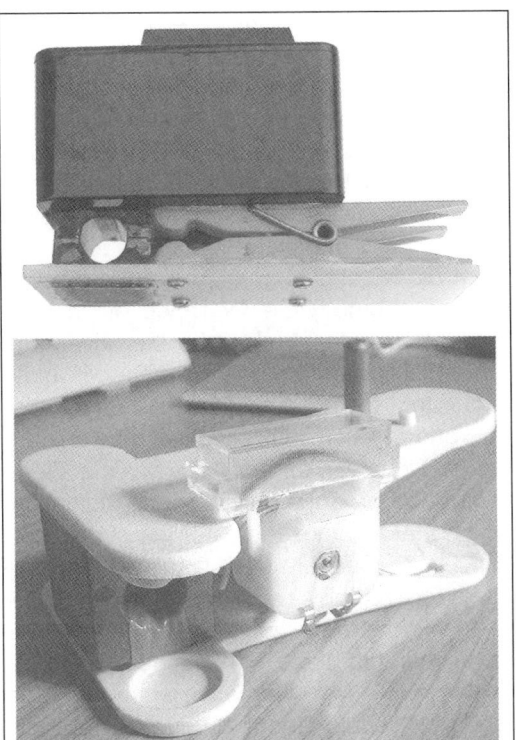

Pic 8.17 Two versions of the clip-on RF current meter

Another option worth investigating would be the pliers-style plastic work clamps that are sold in a range of sizes by hobby shops. Whatever you use, it is vital that you glue the two halves of the core to the clip in such a way that they always close tightly together with no air gap. Hint: glue one half of the core to one side of the clip first, and let that side set; don't try to glue the second half until the first is good and solid.

A clip-on RF current meter could hardly be simpler to build and is an ideal project for beginners and clubs.

→ [1] This article (with earlier references) is from 'In Practice', by Ian White, GM3SEK, *RadCom*, December, 2003.

→ See, 'Clamp-on RF Current Meter - The most useful tool for RF interference troubleshooting!' on GM3SEK's 'In Practice' website. Design based on G0SNO's original article in *RadCom*, April 1993, page 74. *http://*

www.ifwtech.co.uk/g3sek/clamp-on/clamp-on.htm

→ See also: 'Clip-on RF Current Probe', David Lauder, G0SNO, *RadCom*, October 2016.

8.12 RF Power Meters

Broadband RF Power Meters

Although it may be hard to find, the low cost Analog Devices AD-8307 (8-pin DIL or SMD) IC provides a good workhorse for making an RF power meter. This IC will accurately measure power levels from -75dBm to +15dBm (a range of 90dB), over a frequency range from DC to in excess of 500MHz. The device is a logarithmic amplifier and responds to voltage, not power, but this is not a problem, as the voltage across a dummy load can be measured and converted to power. The output voltage is DC and scaled at 25mV/dB which is generated by a current of nominally 2mA/dB through an internal 12.5kΩ resistor. This voltage varies from 0.25V at an input of -74dBm up to 2.5V for an input of +16 dBm. This slope and intercept can be trimmed using external adjustments. With a 2.7V supply, the output scaling may be reduced, for example to 15mV/dB, which permits utilisation of the full dynamic range. For further details see the Analog Devices data sheet. The data sheet even has a suggestion for a 1µW to 1kW power meter (90dB range). The meter can be a typical analogue type, or alternatively a digital voltmeter.

→ There have been quite a few projects based on this chip; see:

'Digital Wattmeter', designed by Thomas Scherrer, OZ2CPU,

http://www.webx.dk/oz2cpu/radios/miliwatt.htm

'Portable RF Sniffer and Power Meter' by Paul Wade, W1GHZ,

http://www.w1ghz.org/new/portable_powermeter.pdf

'Simple RF Power Measurement' by Wes Hayward, W7ZOI, and Bob Larkin, W7PUA.

http://w7zoi.net/Power%20meter%20updates.pdf

→ Note that RF power meter kits are also available, for example:

Fox Delta products PM5 'Dual Channel 500MHz RF Meter with 30MHz Frequency Counter'. A project by Tony/I2TZK, Frank/K7SFN, and Dinesh/VU2FD.

http://www.foxdelta.com/products/pm5.htm#buy

'USB 0-500MHz RF Power Meter with AD8307'

http://electronics-diy.com/AD8307_USB_0-500MHz_RF_Power_Meter.php

→ A watt-meter and swr-meter design using dual ferrites developed by Dave Stockton G4ZNQ based on a broadband coupler in the 1996 *ARRL Handbook* pp 22.36 can be seen at:

www.sm7ucz.se/Meters/Stockton_pwr_meter.pdf

Kits based on the 'Stockton' wattmeter may be available via The North Georgia (NoGa) QRP Club at a cost of around £15. See NoGaWaTT Kit Information at:

http://www.nogaqrp.org/projects/NOGAwatt/nogawattwithnewschematic.doc

A Simple Sensitive RF Milliwatt Meter

The following is an article written by Joachim Köppen, DF3GJ, and published in *RadCom*, February 2008.

For home-brew projects one often wants to know the RF output power of an oscillator or an amplifier stage, for instance to be sure that a mixer is driven with a sufficient level. So we need to measure powers of a few milliwatts or less.

A simple RF 'diode sniffer' – where the RF voltage is rectified by a diode and the resulting direct current or voltage is measured – is fine for powers of more than a few milliwatts. This is because a signal of 10mW across a load of 50Ω has a peak amplitude of 1V, which is sufficient to overcome the typical 0.2V forward threshold voltage of a germanium diode, and to give a useful indication on (for instance) the 1V range of an analogue multimeter. It also has the advantage that the rectified voltage corresponds quite accurately to the amplitude of the RF signal, and we can thus measure the signal's power. However, for smaller signals a 'diode sniffer' will produce only a feeble indication, which makes adjustments of the circuit a bit difficult. To measure such small signals, one could forward bias the diode slightly in order to lower the detection threshold; at these low levels, the diode works in its square-law characteristic. An alternative approach is to load the diode's output as little as possible: Experiments with a simple diode rectifier circuit showed that the rectified voltage depends quite strongly on the load resistance presented by the voltmeter. A standard 20kΩ per volt multimeter gives higher readings on the 2.5V range than on the 1V range; also, a digital multimeter always gives higher readings than the analogue meter. DF3GJ investigated this a bit more systematically with an

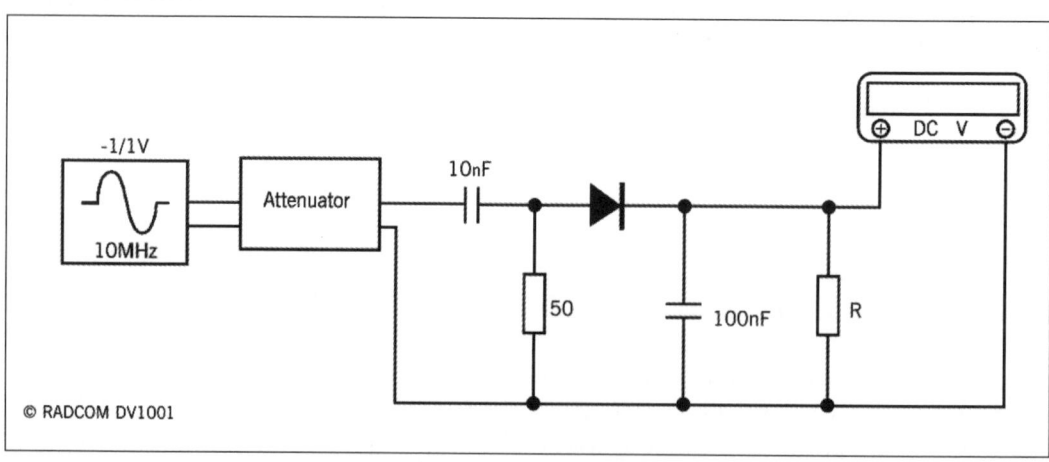

© RADCOM DV1001

Fig 8.70 Test circuit to measure the rectified voltage as a function of the RF power and the value (R) of the load resistor

oscillator giving +8 dBm (6.3mW) at 10MHz, a home-made step attenuator, a 50Ω load, and a germanium diode (see **Fig 8.70**). The results are summarized in **Fig 8.71**.

If it was an ideal rectifier, the rectified voltage would be equal to the RF amplitude, as the smoothing capacitor is charged up to the peak value. With a real diode it is a somewhat smaller output voltage. At levels larger than about -10dBm (100µW) this is but a small error, and the output voltage increases with increased power as one would expect: the diode works in the linear region. However, for smaller signals there is an increasing deviation from the ideal value; for instance, a signal of -22dBm (6.3µW) has an amplitude of 26mV but only 8mV is observed. This is the value measured with a high input resistance digital multimeter which puts essentially no load on the diode. With a load resistance of 15kΩ, typical for an analogue multimeter, a mere 1.6mV is observed, which is less than the pointer's width! Here the diode works with a square-law characteristic - the output voltage depends on the square of the RF amplitude. Note that the curve with no load is steeper than the one from the ideal rectifier. The broken line in the figure shows an ideal square-law response. Hence between -22 and -12dBm this diode is very close to a square-law device.

With a silicon diode one finds that the resulting rectified voltages are somewhat smaller. On the other hand, a Schottky diode gives a

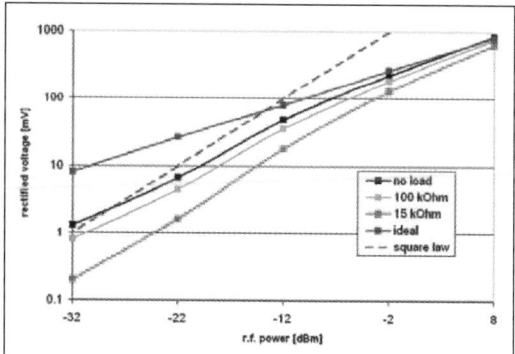

Fig 8.71 Measured voltage rectified by a germanium diode for various values of the RF power and the load resistance

somewhat higher voltage than a germanium diode. In all cases, the voltages go down with lower load resistance.

One option would be to measure the rectified voltage with a digital meter. If it is provided also with an analogue indicator, eg a bar-graph to show changes of signal level, this diode/meter combination could also be useful for alignment and tuning.

Using an analogue meter

By using a standard analogue meter of, say, 100µA and a couple of transistors one can construct a useful and sensitive standalone instrument. The circuit is shown in **Fig 8.72**. The RF voltage across the 50Ω termination resistor (either a single 51Ω or two 100Ω units in parallel) is rectified by two germanium or Schottky diodes, to get both a positive and a negative voltage. The 10MΩ resistor R4 ensures a slight forward bias for diode D1. If germanium diodes are used, this resistor may not be necessary, since the leakage of these will give the necessary bias. Because of their much lower leakage, silicon or Schottky diodes would not be properly biased without R4. While any of these could be used, silicon diodes – eg the 1N4148 – give a lower output, but Schottky diodes give a slightly higher output than germanium diodes. It is useful to select a pair giving the same forward voltages.

The author used the almost obsolete germanium diodes because there are many in his junk box. The output current from each diode is amplified by two emitter followers connected as a Darlington pair. This combination of two transistors gives an input resistance equal to the emitter resistor multiplied by the product of the transistors' DC gain (ß - approximately 200), hence about 40MΩ. Each amplifier has unity voltage gain, and thus the difference of the emitter voltages can be measured with a normal 100µA moving coil meter to give an indication of the RF amplitude.

The balancing trick

The two resistors R2 and R3 bias the transistors to have their input at about half the

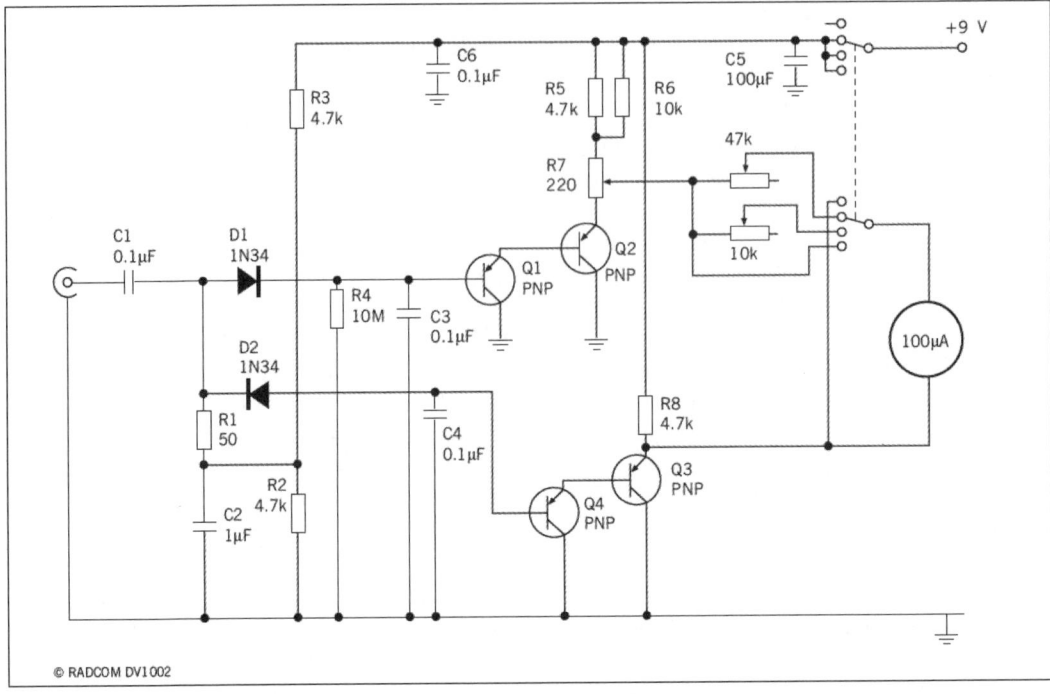

© RADCOM DV1002

Fig 8.72 The circuit diagram of the milliwatt meter

supply voltage, hence the emitters of the second transistors are about 1.4V above that level. Since the diodes and transistors are not exactly equal, and their characteristics also change with the ambient temperature and battery voltage, it is necessary to adjust the balance of the instrument to get a zero indication for a zero input signal. This control is provided by inserting a small potentiometer into the emitter branch of one of the amplifiers. It is best determined by experimentation into which side the potentiometer should go, and the values of the potentiometer and resistor R6. The zero adjustment is not critical; done once, sometime after switch-on, the instrument remains well-balanced for a couple of hours. The transistors are general-purpose silicon PNP types from the junk box, 2N3906 or similar – they were not selected for equal DC gain. NPN types, like 2N2222, 2N3904, BC 107 can be used and then one needs to reverse the polarity of the supply voltage and the diodes. Current consumption is quite low, drawing a few milliamps from a small 9V block.

As seen in the photograph of **Pic 8.18**, the instrument is built into a case constructed from single-sided printed circuit board. The copper side serves as ground and shielding. All components are soldered to this board, without any special precaution for the arrangement, except that RF paths remain short. In particular, C1 and C2 should be low inductance types with very short leads (eg surface mount). I chose to have three ranges of sen-

Pic 8.18 The interior of the finished instrument. From left to right are the range switch with trimmer resistors, the coaxial input socket, the diodes and the transistors, and the zero adjustment potentiometer

sitivity; for +10, 0 and -10dBm at full scale deflection. This gives a large overall range of about 30dB with generous overlap, which is useful for alignment operations.

Calibration

The instrument requires calibration on each range. This is done with an oscillator of about 10mW output and a step attenuator. The latter can be constructed for an impedance of 50Ω with carbon film or metal film resistors. Determine the maximum output with a circuit as in **Fig 8.70**. Load the oscillator with 50Ω and, using a germanium or Schottky diode, measure the rectified voltage with a digital voltmeter. Since the peak voltage is measured, the power is computed via P=V2/2R, which gives 10mW for an amplitude of 1V. Use various attenuator settings to establish a calibration table for each range, or calibrate the scales on the dial directly. The dials were calibrated at 5MHz, but although no special precautions were taken in the layout of the circuit the finished instrument gives satisfactory results at 144MHz and even at 450MHz - as checked with the aid of professional attenuators. Also, switching over to another range gives consistent readings. **Pic 8.19** shows the measurement of the oscillator of a TV UHF tuner.

Summary

This has proved to be a handy and often used instrument which allows one to measure signals as low as -20dBm (10µW) and positively detect signals down to -25dBm or somewhat less. Thus, the output from small signal amplifiers and frequency multiplier stages could be picked up easily, even when they are still detuned and not yet optimised. In conjunction with the calibrated step attenuator, it allows one to measure the gain of small signal amplifiers as well as to determine their saturation level. It takes much of the guesswork out of home-brewing.

QRP Wattmeter

The wattmeter described here will read up to a maximum of 3W and is usable up to a frequency of 30MHz. It is in essence a peak reading voltmeter with internal 50Ω dummy load and is not designed to read standing wave ratios. Sufficient information is given for the design to cope with varying meters and full scale power levels.

Circuit description

The circuit of the complete unit is given in **Fig 8.73**. Resistor R1 forms the dummy load; D1 provides rectification; C1 smoothing; R2 limits the current through the meter M, and C2 provides RF decoupling.

The dummy load should be made from carbon resistors and of adequate rating to cope with 5W. As a minimum, use 3x270Ω and 2x220Ω 1W resistors to give a load of 49.5Ω. Another arrangement could be 4x330Ω and 3x390Ω, giving an equivalent resistance of 50.5Ω. Other arrangements are also possible.

The diode needs a little consideration, but it must be capable of high speed working, have a minimum PIV of 23V, and as low a forward-drop as possible in order to minimise errors at the low power end. Although

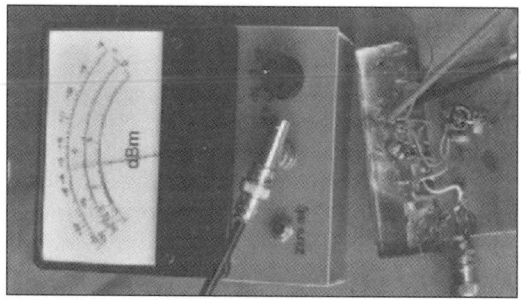

Pic 8.19 The milliwatt meter in use: measuring the output of a UHF oscillator from a TV tuner

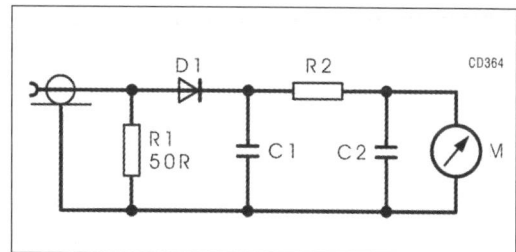

Fig 8.73 Circuit diagram of QRP wattmeter

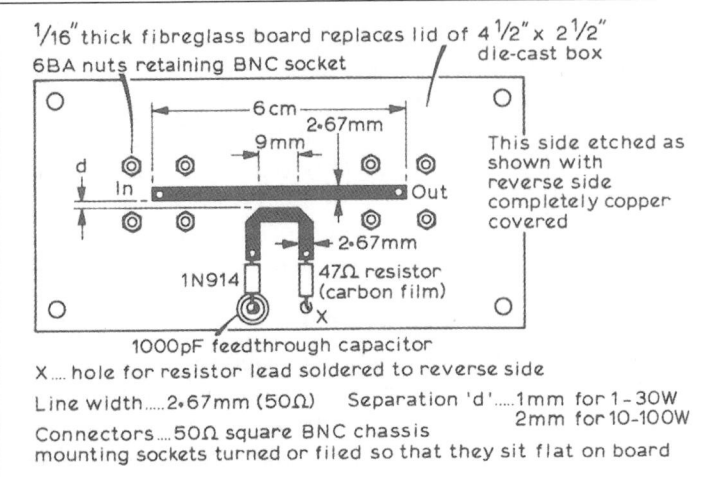

Under 0.1W the forward voltage-drop of the diode becomes significant.

Fig 8.74 A simple forward power indicator for a 50 ohm line

the ubiquitous 1N914/1N4148 is more than adequate, a lower forward voltage-drop can be obtained from a Schottky diode such as the BAT85. However, for higher power levels use the 1N914. Both capacitors should be of the ceramic type, with at least a 30V rating. Resistor R2 can be calculated to cope with various sensitivity meters. Use a meter of between 50µA and 1mA sensitivity. Neglecting the meter resistance, R2 is given by:

R2 = V/I$_{fsd}$

where V is the peak value of the rectified sinewave, Ifsd is the sensitivity of the meter, and V is calculated from:-

V=10√P, where P is the power being measured.

Thus for a 100µA meter and requiring full scale deflection for 5W; R=223.61kΩ (220kΩ+3k6). The meter resistance of about 1kΩ is negligible compared with this. The current corresponding to a given power is given in the table below.

(W)	(uA)
0.1	14
0.5	32
1.0	45
2.5	71
5.0	100

Construction

It is suggested that the whole unit is mounted in a metal box with some ventilation for the dummy load. The circuit from diode to meter should be kept as far away as possible from any circuits carrying RF, and shielded if at all possible. Use a suitable connector.

1296MHz In-Line Power Indicator

A simple and reliable RF power indicator for insertion into the output line of a 1296MHz transmitter can be readily constructed by taking advantage of microstrip techniques. For this purpose, good quality glass-fibre double-sided board is needed (eg FR4). Leave one side as an earth plane and etch the other side as per **Fig 8.74**.

The insertion loss of this type of indicator is of the order of 0.5dB and it may therefore be permanently connected in circuit. The spacing between the line and the coupling loop will need to be decided on the basis of the power expected to be normally used (ie voltage on the line). The whole assembly should be enclosed in a metal box.

Although the device is defined as a forward indicator, if the connections are reversed, it may alternatively be used for indicating reflected power.

Other Projects

→ See 'A Precision Peak-Following Power Meter' by Brian Horsfall, G3GKG, *RadCom*, March 2001. This meter covers the HF bands, coping with powers up to about 450W and requires no setting up or adjustment during use. A finished unit is shown in **Pic 8.20**. An abridged description is in the

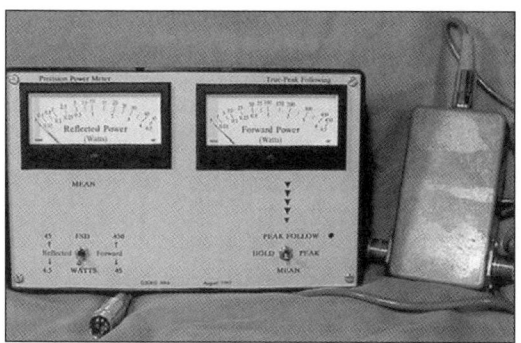

Pic 8.20 Precision peak-following power meter

previous edition of this book: *Test Equipment for the Radio Amateur*, (4th Edn), Clive Smith, G4FZH RSGB.

→ See also 'The Crawley Power Meter' by Derek Atter, G3GRO, and Stewart Bryant, G3YSX: *https://archive.org/stream/TheCrawley-PowerMeter/The_Crawley_Power_Meter#page/n4/mode/1up*

8.13 Modulation Meters

A Modulation Meter for SSB

The level of modulation of an SSB transmitter is usually judged by the movements of a moving coil meter; however, a snag immediately arises because the meter, due to its inertia, cannot respond quickly enough to the speech waveform. The modulation will reach a peak level and decay before the meter needle reaches the peak current value.

The advice usually given, is to limit the indicated current on speech peaks to half that

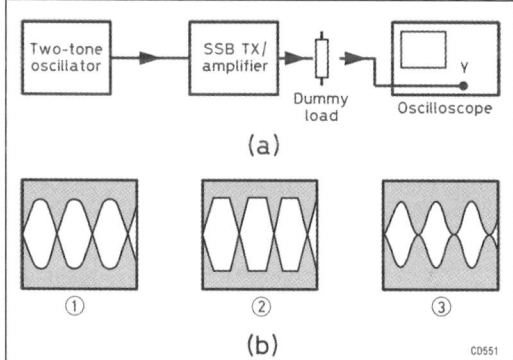

Fig 8.75 Two-tone amplifier/transmitter testing

determined by the two-tone test. **Fig 8.75** shows the test arrangement and the range of traces obtained from an audio two-tone sinewave, fed in at the microphone socket.

These observations are made under steady state conditions, and from them, the current which should not be exceeded when speaking into the microphone can be deduced. However, when it comes to observing the limit (of 50% of this) on the transmitter's meter, the advice may be somewhat difficult to implement. Meters vary in their response to spiky waveforms, some being heavily damped and others extremely responsive. In the latter case, mechanical resonances come into play as well and the needle can dance all over the place and give no real guide as to what is happening. It was decided that an indicator in the form of an LED bargraph would solve the meter response problem. The equipment described consists of an RF detector and a display unit.

Circuit description

The circuit diagram for the display unit is shown in **Fig 8.76** with the component listing in **Table 8.13**. IC1 is a bargraph driver which also contains comparator circuits. Ten individual LEDs are used but it is possible to buy 10 LEDs in a single module. The nine LEDs for the lower modulation levels are green, the tenth LED, which gives a warning of excess

R1	1k2
R2	12k
R3	10k
RV1	4M7
C1	10n, 50V ceramic
C2	1n, 50V ceramic
C3	47n, 100V ceramic
C4	2n2, 160V
D1	OA91 or similar
LEDs	Separate or module
IC1	LM3914
IC2	7805, 5V, 500mA
3.5mm jacks and socket	
Connectors as appropriate	
Resistors are 0.25W/0.5W, 5% unless specified otherwise.	

Table 8.13 Component listing for SSB modulation meter

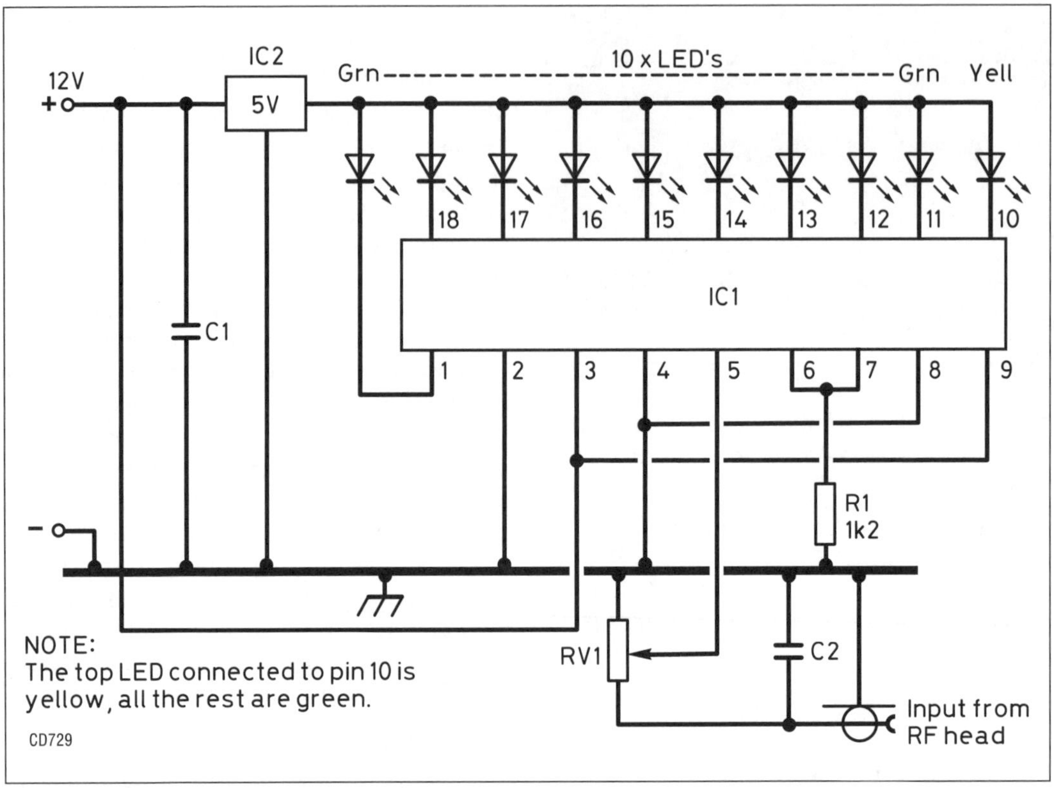

Fig 8.76 Circuit of the bargraph unit

modulation, is yellow. The internal reference level is used for the internal comparator chain and all the LEDs will illuminate for an input of 1.25V or greater (pin 5). The level from the RF detector head is adjusted by RV1 so that for a given transmitter nine LEDs illuminate for maximum modulation. By providing RV1 with a scale (dial) settings may be determined for other transmitters or linear amplifiers having different power ratings. The rise-time of the circuit for a step input is determined by R2 and R3 in parallel, and C3. Thus the combination of a resistance of 5450Ω and a capacitance of 47nF gives a time constant of 256µs. In order to show up short modulation peaks of a few milliseconds, the fall time needs to be greater than, say, 100ms. This is determined by RV1 and C3 and produces a time constant of the order of 220ms.

The display unit requires a DC supply, so for this application, 12V is fed to pins 3 and 9, the LEDs being fed with 5V at a typical maximum current of 150mA supplied by IC1. Resistor R1 determines the LED currents and C1 provides power supply decoupling. Capacitor C2 provides additional smoothing for the detected RF. Construction of the display unit is not critical.

The circuit of the detector head is shown

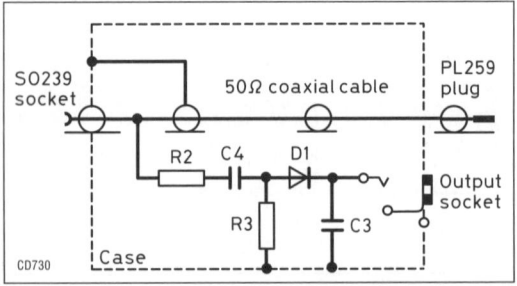

Fig 8.77 Circuit of the RF detector head

in **Fig 8.77**. The unit provides a tapping on the 50Ω coaxial antenna lead from which a sample of the RF is rectified and smoothed for driving the bargraph display. The signal from the potential divider formed by R2/R3 is detected by diode D1, C4 providing DC isolation and C3 the smoothing. Note: some transceivers have DC on the antenna lead for control of linear amplifiers hence the requirement for C4.

Construction

The display unit was mounted in an aluminium case 100x65x50mm. The PCB can be mounted so that the potentiometer spindle passes through the case and a dial arrangement can be attached. Also provided is a

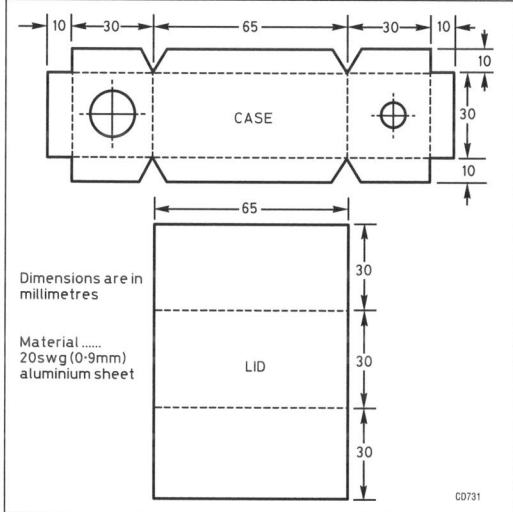

Fig 8.78 Details of case for RF detector

3.5mm jack socket for the detector head and power supply sockets.

The detector head is housed in an aluminium enclosure 65x30x30mm, details of which are given in **Fig 8.78** - this could also serve as a pattern for the display unit box with suitable dimensions. The ends of the box need to be just large enough for a connector socket on either end. **Fig 8.79** shows the layout of components in the detector head.

In use

The separate detector head allows the display to be located in any convenient position, while the head can be inserted at any suitable point in the coaxial antenna lead, providing it is monitoring the correct signal.

Calibration of the bargraph input permits the monitoring of different power levels. In the prototype, R2 and R3 were adjusted so that 3W PEP from a 144MHz transceiver just lit nine LEDs at a setting of RV1 near 100 on the dial. The lower scale readings for 25W and 100W PEP outputs were then recorded using an oscilloscope to indicate the onset of flat topping. The component values are such that the unit should work satisfactorily down to 1.8MHz.

Choosing values for R2 and R3

The unit was designed for a maximum transmitted power of 100W PEP. For higher powers the potential divider R2/R3 needs to be adjusted to ensure that the OA91 is not subjected to a PIV approaching 115V.

1. Find the RMS voltage on the 50Ω line: $V=\sqrt{(50)W}$ where W is the transmitter output in watts.

2. If V_2 is the RMS voltage at the diode anode, then: $V_2 = V_1R3/(R2+R3)$

3. Calculate the PIV for the diode $=2.83 \times V_2$. Ensure that this is well within its rating.

4. Check the wattage ratings of R2 and R3: Power dissipated in $R2=(V_1-V_2)^2/R2$; Power dissipated in $R3=V_2^2/R3$

The power rating for each of these resistors is normally limited to about 0.5W each.

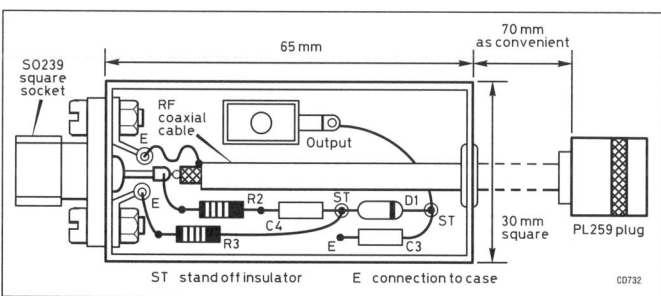

Fig 8.79 Arrangement of components in detector head

→ This modulation meter is by John Stebbings, G4BTV, and appeared in Radio Communication, March 1985.

8.14 Wobbulators

A sweep generator or wobbulator is an oscillator whose frequency can be varied by, say, the application of a suitable voltage - in essence a voltage controlled oscillator (VCO). If a control voltage is used which varies between two limits (eg a ramp), then the frequency of the sweep generator can be changed or swept between these two limits. They are a useful instrument for aligning such items as IF strips, band pass filters, and setting up FM discriminators, as the response of the circuit can be displayed on an oscilloscope and the effect of tuning adjustments can be seen immediately without recourse to step by step methods. The variation in frequency is normally achieved by varying a supply voltage, or by using a varicap diode with an LC oscillator.

A prime requirement of a sweep frequency generator is that it should be completely free of any amplitude modulation over the range to be swept and have a linear change of frequency with sweep voltage. While these features can be achieved by careful design, simpler devices all too frequently have considerable amplitude modulation. If this defect is not minimised, it is possible to produce response curves on the oscilloscope which appear to be ideal whilst the true response of the equipment being aligned might be quite unsatisfactory.

Practical use of sweep generators

Fig 8.80 shows the typical arrangement for

displaying the frequency characteristics of a circuit under test. The ramp signal which is used to sweep the RF oscillator is also applied to the oscilloscope X input (ie internal timebase switched off). The output of the cir-

R1	1k2
R2	10k
R3	2k
R4	100k
R5,6	10k
R7	39k
R8	47R
R9,10	1k
RV1	22k lin
RV2	10k log
C1	1µF
C2	100n
C3	10n
C4,5,6	100µ, 25V
C7	47µF, 25v, non-polarised
C8	1n
IC1	NE566
IC2	741
SW1	1p, 12w
Resistors are 0.25W/0.5W, 5% unless specified otherwise.	

Table 8.14 Component list for triangle-wave generator

cuit under test is fed to the oscilloscope Y input. The gain controls on the oscilloscope can then be set to give a typical display as shown, thus directly displaying the overall frequency response curve of the circuit under test.

What is required, ideally, is a storage oscilloscope – or possibly the computer equivalent – with an X-input driven by an external signal source, the frequency of which is swept very slowly through a variable portion of the appropriate spectrum. Auto-synchronisation of the display is then achieved because the X-deflection of the oscilloscope is the linear ramp voltage which is causing the frequency deviation. With a digital storage oscilloscope that doesn't allow storage when plotting X against Y, the

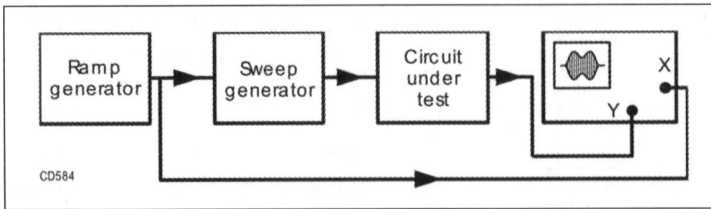

Fig 8.80 Connections for oscilloscope display

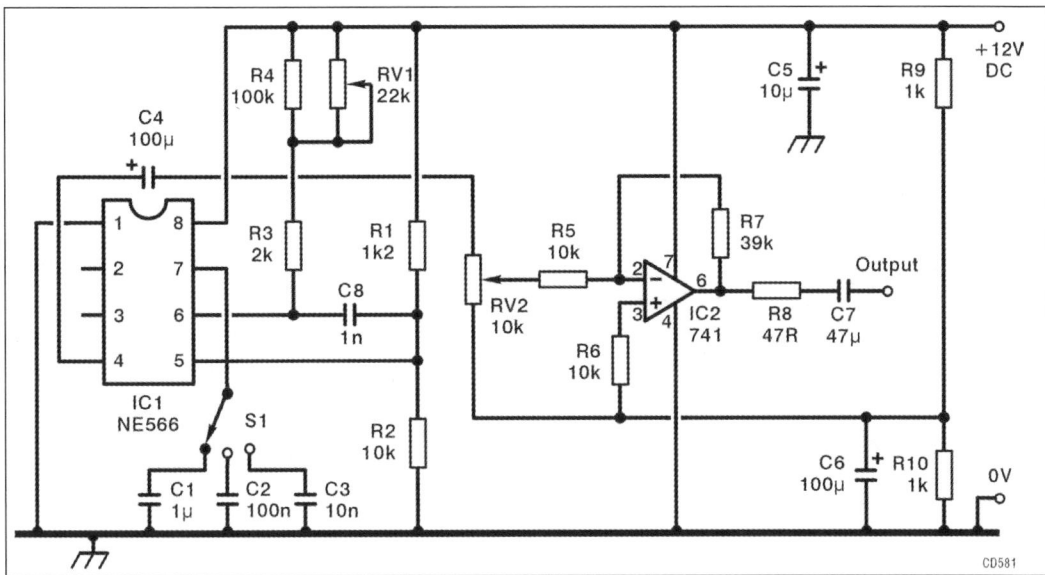

Fig 8.81 Circuit diagram of triangle wave generator

setting up procedure is rather more complicated but, for that situation, this design does provide an output trigger pulse which permits manual synchronisation of the oscilloscope's own time base.

A Triangle-Wave Generator

This triangle-wave generator can be used to drive the two sweep generators, one for each of the two most common IF frequencies described later.

The circuit is shown in **Fig 8.81** with a component listing in **Table 8.14**. It formed around the NE566 function generator (IC1), which is a voltage controlled oscillator. The output is a triangle waveform of high linearity and about 2.5V peak to peak. This is then AC coupled to a 741 amplifier (IC2) with a gain of just under 4. The potentiometer RV2 allows the output to be varied between zero and 10V peak to peak. The frequency is controlled by SW1 and RV1. There is also a square-wave available at pin 3 of IC1. The ranges are approximately:

Range 1: 10Hz–100Hz

Range 2: 100Hz–1kHz

Range 3: 1kHz–10kHz

Construction of the circuit is not complex but a PCB and component layout is given in Appendix D.

440-550kHz Sweep Generator

Generally there is no need for a sweep generator to produce sine waves: a square-wave generator of suitable repetition frequency is equally effective. The basic multi-vibrator may be readily modulated in frequency with a linear sweep while giving an output of constant amplitude; this is the basis of the design shown in **Fig 8.82** and described here. For a component listing see **Table 8.15**. The values given are suitable for 440-450kHz, but with suitable component changes the circuit can cope with centre sweep frequencies in excess of 20MHz.

Circuit description

The centre frequency (tuning) is adjusted by varying the voltage applied to the base of TR2 by RV2. The amount the frequency is swept (deviation) is controlled by varying the amplitude of the input sweep voltage via RV1. The sweep voltage is AC coupled to ensure symmetrical deviation. The transistors used should be of medium current gain (ie

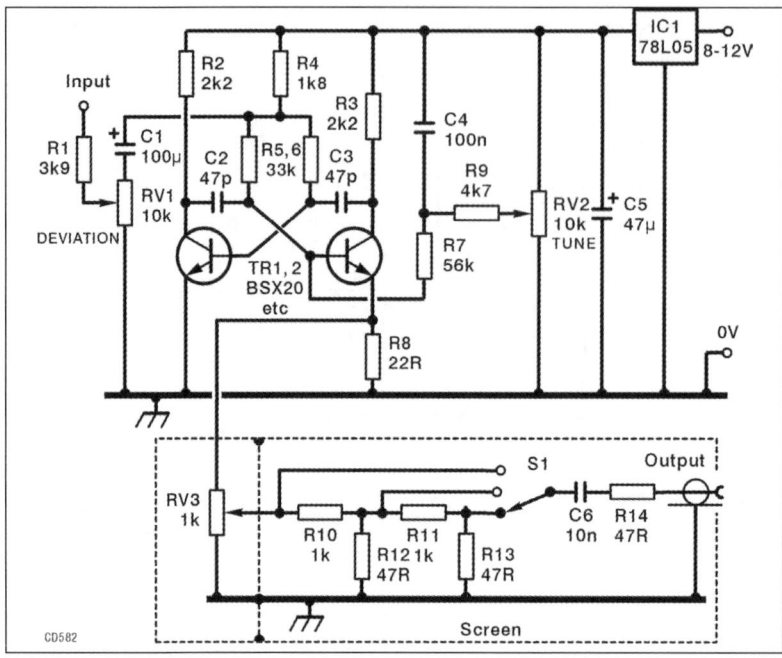

Fig 8.82 Circuit diagram of 440-550kHz sweep generator

30-50); if high gain devices such as a BC109 are used both transistors will turn on together and the circuit will not oscillate.

The output voltage is taken from the emitter

R1	3k9 (see text)
R2,3	2k2
R4	1k8
R5,6	33k
R7	56k
R8	22R
R9	4k7
R10,11	1k
C1	100µF, 16v electrolytic
C2,3	47p
C4	100n, ceramic
C5	47µF, 16v electrolytic
C6	10n, ceramic
TR1,2	BSX20 or similar (see text)
IC1	78L05
SW1	1p, 3w rotary switch
R12,13,14	47R
RV1,2	10k pot
RV3	1k pot.
Resistors are 0.25W/0.5W, 5% unless	
specified otherwise	

Table 8.15 Component list for 440-550kHz sweep generator

of TR2 via potentiometer RV3 and the 3-position attenuator formed around SW1. Provided the output circuit is shielded, the control of output is quite adequate to allow the effects of AGC on change of gain of IF response to be checked. The circuit operates at 5V, which is supplied by IC1, whose input can be in the range 8-12V.

Component layout is not critical, although the attenuators should ideally be screened from the rest of the circuit. A PCB and component layout is given in Appendix D. The tuning controls can be checked by monitoring the output of the unit with a digital frequency counter.

Note: if the sweep signal is obtained from an oscilloscope timebase output, R1 may need to be included. This is estimated at about 40kΩ per 10V pk-pk of time-base voltage.

A 9 or 10.7MHz sweep generator

This design is shown in **Fig 8.83** with the component list in **Table 8.16**. It is based on a Clapp oscillator formed around TR1. This is then followed by a buffer amplifier TR2. The centre frequency is set by the variable capacitor VC1 and the slug in the coil. The values given will adjust between 9 and 11MHz but could be modified for other frequency ranges.

The sweep signal is applied to a variable capacitance diode D1 and the swept frequency adjusted by RV1 in conjunction with R1. If an oscilloscope timebase is used, the value for R1 may need to be adjusted.

The deviation can be calibrated by applying DC voltages to the diode and observing the

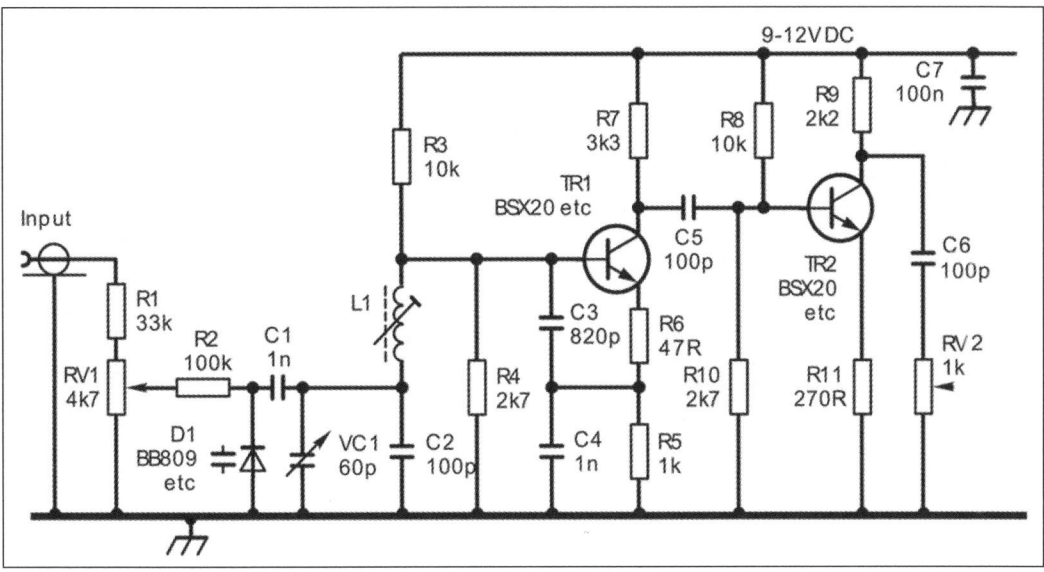

Fig 8.83 Circuit diagram of 9-10.7MHz sweep generator

output frequency on a digital frequency meter. The amplitude of the output is controlled by RV2, the maximum output being about

R1	33k (see text)
R2	100k
R3	10k
R4	2k7
R5	1k
R6	47R
R7	3k3
R8	10k
R9	2k2
R10	2k7
R11	270R
RV1	4k7 pot.
RV2	1k pot.
C1	1n, polystyrene
C2	100p, polystyrene
C3	820p, polystyrene
C4	1n, polystyrene
C5,6	100p, ceramic
C7	100n, ceramic
VC1	60p, e.g. Jackson C804
TR1,2	BSX20 or similar
D1	BB809 or similar
L1	TOKO 113KN2K1026HM; or 12t, 26swg on 7mm former with iron slug, 1.4µH

Resistors are 0.25w/0.5W, 5% unless specified otherwise.

Table 8.16 Component list for 9-10.7MHz sweep generator

500mV (RMS). A sweep of up to 5% of the centre frequency (about 0.5MHz) can be obtained without significant amplitude change. A PCB and component layout is given in Appendix D.

→ To measure crystal and other high-Q, steep-sided filters, see 'A Slow Scan Wobbulator' by Brian Horsfall, G3GKG, *RadCom*, May 2005, p85ff, or the abridged version of this article in a previous edition of this book: *Test Equipment for the Radio Amateur*, (4th Edn), Clive Smith, G4FZH RSGB.

8.15 Power Supply Units

Fixed 9v, Low Current Power Supply

This circuit can replace the familiar PP3 battery which is used in many items of portable test equipment. The circuit will supply 9V at up to about 140-150mA and uses the ubiquitous 723 regulator IC which has been around for many years and has short circuit current protection. **Fig 8.84** shows the circuit diagram for this unit and the component list is in **Table 8.17**. A sample of the output voltage (provided by R1/R2) is compared with an internal reference, whilst R4 is the current sense resistor which limits the output current. A PCB and

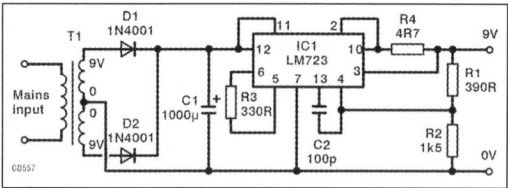

Fig 8.84 Fixed 9V, low-current supply

R1	390R, 1%
R2	1k5, 1%
R3	330R
R4	4R7
C1	1000µF, 25V electrolytic
C2	100pF, 50V Ceramic
T1	240/9-0-9V, 3VA
D1,2	1N4001 or similar
IC1	LM723 or equivalent
Resistors are 0.25W/0.5W, 5% unless specified otherwise	

Table 8.17 Component list for the fixed 9V, low-current power supply

component layout is given in Appendix D.

Mid-Point Power Supply

Sometimes it is necessary to generate a mid-point voltage in order to create complementary supplies. It is possible to use just a potential divider using equal resistors but this will have a low current capability. The circuit shown in **Fig 8.85** uses a 741 operational amplifier (or equivalent) and will source or sink 25mA. The output capability can be further increased by the use of an output buffer.

Fixed 5V/12V, 1.5A+5A Power Supply

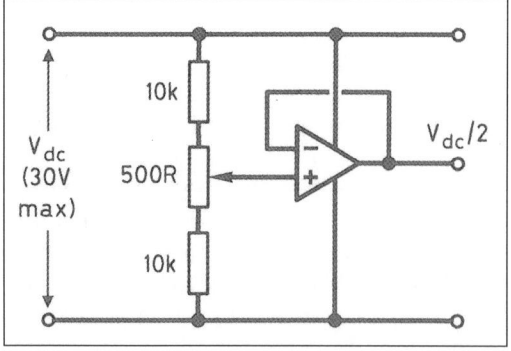

Fig 8.85 Mid-point power supply

These power supplies use fairly standard components and the circuits are the same except for component values. Because of this, a single design is presented in **Fig 8.86**. In both the 1.5A and 5A supplies it is suggested that the bridge rectifier is bolted to the metal chassis. A universal component list is given in **Table 8.18**; however, other bridge rectifiers and regulators exist that will also work fine. The regulators should be placed on heatsinks using heatsink compound. The minimum suggested heatsink

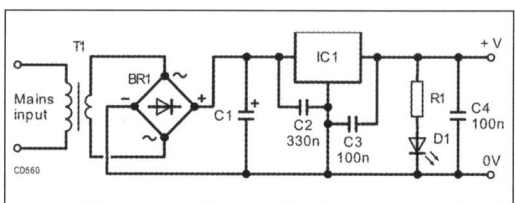

Fig 8.86 Fixed output voltage, regulated supply

values are:

5V, 1.5A:	2.5°C/W
5V, 5A:	1°C/W
12V, 1.5A:	1.8°C/W
12V, 5A:	0.8°C/W

C2 and C3 should be placed as close to the regulator as possible, whilst if C4 is included, it should be placed across the output terminals of the power supply.

A Versatile ±12V, 1.5A PSU

In certain circumstances, a power supply is required with equal positive and negative voltages. One way of achieving this is to use the circuit in the previous section twice, but using a single mains transformer with dual output windings which probably represents the most versatile arrangement. **Fig 8.87** shows the general arrangement with individual outputs AB and CD, with no terminal grounded. There are thus initially two independent 12V outputs, however a ground can be applied separately to one of the output terminals of each supply. If B and C are connected there is then 24V between terminals A and D. If the junction BC is grounded,

Output	Transformer	BR1	C1	IC1	R1
5V, 1.5A	9V, 2.7A, 30VA	GBU4B	4700µF	L7805CV	220R, 0.25W
5V, 5A	9V, 8A, 80VA	GBPC12005	15000µF	LT1084CT5	220R, 0.25W
12V, 1.5A	15V, 2.5A, 30VA	GBU4B	4700µF	L7812CV	680R, 0.25W
12V, 5A	15V, 8A, 120VA	GBPC12005	15000µF	LT1084CT12	680R, 0.25W
For all versions C2 is 0.33µF, 63V, Polyester; C3, C4 are100nF, 50V, Ceramic and D1 is a Red LED					

Table 8.18 Component list for the fixed 5V or 12V, 1.5A and 5A power supplies

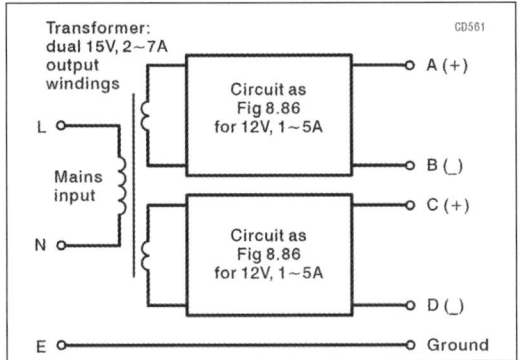

Fig 8.87 General arrangement for a versatile ±12V, 1.5A power supply

A is 12V above ground and D is 12V below ground, ie at -12V. If A is grounded D is at -24V, if D is grounded then A is at +24V.

Variable Voltage 2.5V-15V, 1.5A Power Supply

This circuit provides a variable output voltage between 2.5V and 15V with 1.5A available at all voltage settings. It uses what is known as a variable voltage reg-ulator. The circuit diagram is shown in **Fig 8.88** and the component listing in **Table 8.19**. Part of the output voltage is fed back via a potential divider formed by R1/ RV1/R2. This feedback ratio determines the output voltage.

In order to optimise load regulation, the resistor R1 should be as close to IC1 as possible whilst the ground end of R2 should be soldered as close as possible to the output terminal. Capacitors C2, C3, and C4 aid ripple rejection and improve transient response.

This arrangement gives a continuously variable output. If only fixed steps are required the

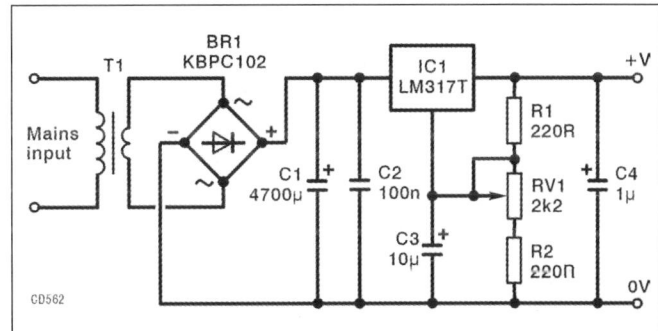

Fig 8.88 A 2.5-15V, 1.5A power supply

R1,R2	220R
RV1	2k2 potentiometer
C1	4700µF, 35V
C2	0.1µF, 50V, ceramic
C3	10µF, 25V
C4	1µF, 25V
T1	240/15V, 2.7A, 40VA
BR1	KBPC102 or similar
IC1	LM317T or equivalent
Resistors are 0.25W/0.5W, 5% unless specified otherwise	

Table 8.19 Component list for the variable-voltage 2.5-15V, 1.5A power supply

resistors RV1+R2 can be replaced by a single switchable resistor R' which is obtained from the following formula: $Vo = 1.25(1+R'/R1)$.

The bridge rectifier should be fixed to a metal surface (eg metal enclosure) and the regulator mounted on a heatsink of 1.5°C/W or better. Note: heat generation is worst at low voltage output. If it is not necessary to go as low as 2.5V the resistor chain can be recalculated and it may then be possible to use a smaller heatsink.

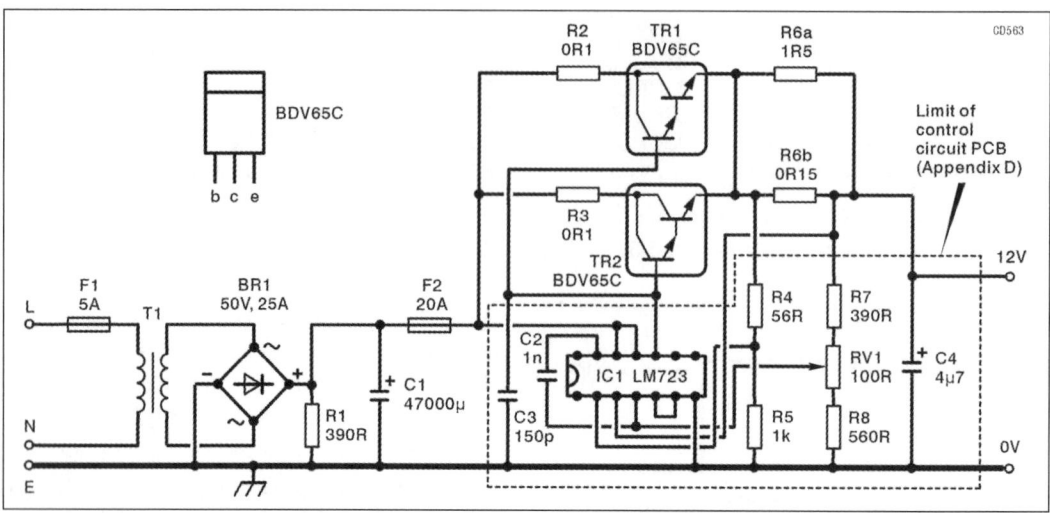

Fig 8.89 Limited-voltage range, nominal 12V, 10A power supply with fold-back current limit

12V, 10A Power Supply With Fold-back Current Limit

Sometimes it is necessary to test equipment with a given input voltage range but high current. For example it may be required to test equipment over the voltage range of a typical 12V vehicle battery or to cope with slightly different voltage specifications. The circuit described will provide a variation from 11.2 to 13.7V.

The circuit diagram is given in **Fig 8.89** with the component list in **Table 8.20**. This circuit uses a low power regulator (IC1) which controls two Darlington pairs (TR1 and TR2). Each of these can carry the total current but the resistors R2 and R3 help to make the devices share the load and the arrangement will give increased reliability. TR1 and TR2 can be spaced apart on the heatsink. Resistors R7, R8 and RV1 form the voltage sampling network. Resistors R6a/R6b form the current sensing network and in conjunction with R4 and R5 provide fold-back current limiting. The variable resistor RV1 will provide the output adjustment and can be either a trimmer or potentiometer. When the output current reaches about 10A (dependent on tolerance of R6a/R6b) fold-back current limit commences and the final short circuit current is 5A. Fuses F1 and F2 should be of a high rupture capacity (HRC) type.

The bridge rectifier and current sense resistors should be mounted onto the metal case the supply is housed in, and the series-pass Darlington pairs TR1 and TR2 mounted (and spaced out) on a heatsink of 0.8°C/W or better. If the power supply is likely to be used for prolonged periods near maximum output, increase the thermal rating of the heatsink and ensure adequate ventilation.

R1	390R,3W, Wirewound
R2,R3	0.1R, 10W, Wirewound
R4	56R
R5	1k
R6a	1R5, 15W, Wirewound
R6b	.15R, 25W, Wirewound
R7	390R
R8	560R
RV1	100R, Cermet Trimmer/pot
C1	47000μF, 25V
C2	1nF, 50V, ceramic
C3	150pF, 50V, ceramic
C4	4μ7, 254V, tantalum
TR1,2	BDV65C
IC1	LM723 or equivalent
T1	240V/15V, 16A, 300VA
BR1	50V, 25A F1 5A, HRC + holder
F2	20A, HRC + holder
Heatsink	0.8°C/W
Resistors are 0.25W/0.5W, 5% unless specified otherwise	

Table 8.20 Component list for limited voltage range 12V, 10A power supply with fold-back current limit

A PCB and component layout are given in Appendix D for the control circuit only. This board can be adapted for other voltages and currents if desired.

A home-brew current sink

Fig 8.90 shows the circuit of the current sink built many years ago by ZS5JF [1] who has updated some of the components (mainly the shunt transistors) for more modern types, and used TIP140 Darlingtons with a current rating of 10A maximum and a collector-emitter voltage of 60V which should cater for most applications. Only two parallel connected shunt transistors are shown as the intention is to rate the current sink at 10A. If more current sinking is required simply increase the number connected in parallel. Do not forget the current sharing resistors (R4 and R5) in each emitter. One could probably substitute N-channel MOSFETs for the NPN Darlington transistors, but this has not been tried.

The equipment requires a DC supply of around 12V and this is regulated with an 8V low power regulator to drive the dual op-amp, an LM358, and the oscillator circuit. This is supplied by a small mains transformer, bridge and electrolytic capacitor or possibly a 'wall-wart', the choice is yours. The variable VR1 needs to be a type that can be reset accurately, eg a multi-turn panel mounting type.

The TIP140 transistors need mounting on a substantial heatsink as they are operating as variable resistors - the same as a series pass regulator - and will generate a lot of heat. The current sharing and sensing resistors need to be adequately rated. The resistors R4, R5, R6, R7 and R8 are chosen for 10A and need to be 5W wire wound types - about 20% of the power will be dissipated in these 'tail resistors'. The polarity protection diode D2 can be eliminated if you are confident you will always connect the power terminals correctly! The rest of the circuit consists of a variable frequency square-wave generator and the metering. This is shown in **Fig 8.91**(a) and (b).

The frequency of the square-wave oscillator is varied by VR2. With the frequency

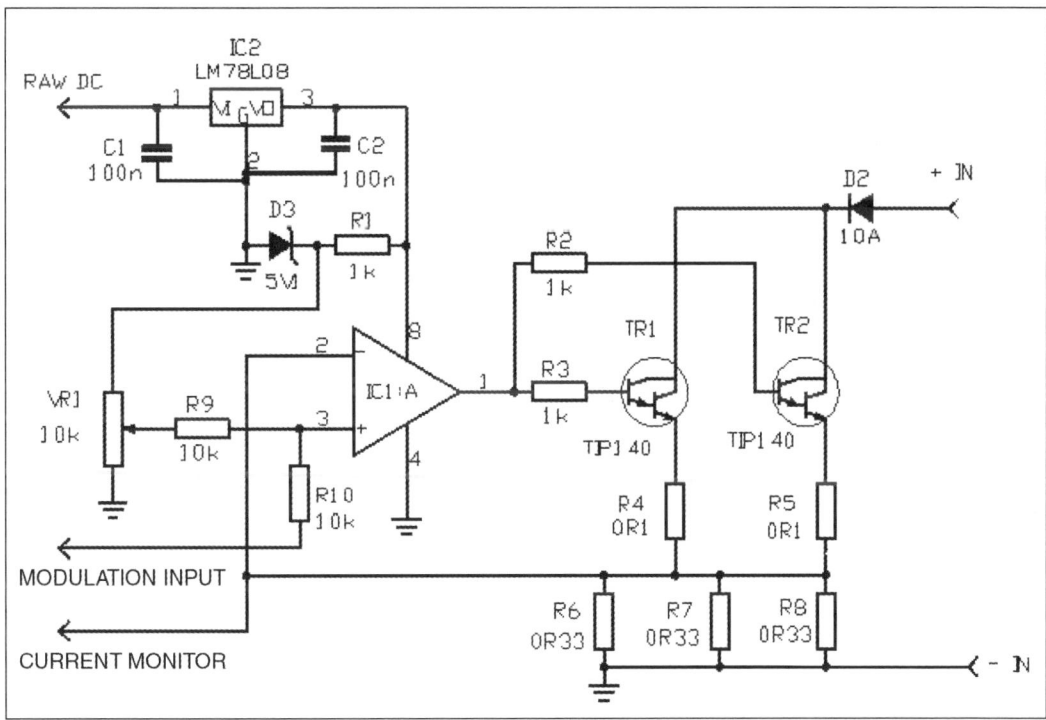

Fig 8.90 Variable current sink power stage

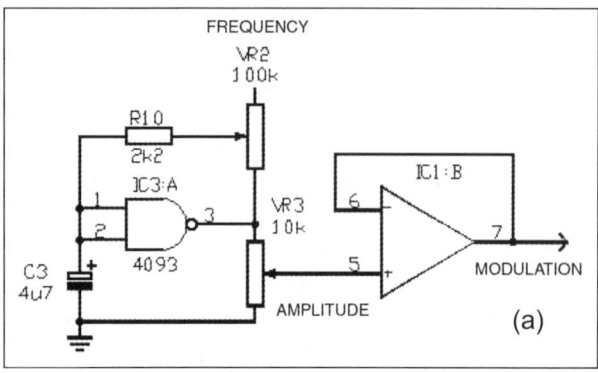 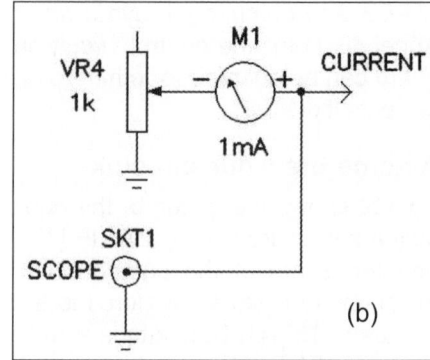

Fig 8.91 (a) Variable current-sink modulator; (b) Current monitor and transient response circuit

control at minimum and the component values shown, the frequency is between about 2Hz and 100Hz on a 50% duty cycle. This range is adequate for SSB and CW. It can be varied over a wider range by changing the capacitor value (C3). VR1 sets the minimum current and VR3 the maximum current. To set up, the minimum current is first set by VR1 (with VR3 at minimum) and then the maximum current is set by VR3. For operation as a constant current sink, the 8V supply to the CD4093 oscillator is switched off. The connector SKT1 allows an oscilloscope to be connected to monitor the current waveform. The preset VR4 sets the current meter to give an accurate indication; adjust it with your DMM in series with the supply to get the correct reading.

The current and voltage transient measurement is taken with a 1mA meter and an oscilloscope.

Additional notes

1. A more elaborate frequency source could be made using one of the multifunction waveform generator ICs, such as an 8038, but the simple square-wave generator works just as well for the sort of power supplies likely to be built and tested.

2. If a shorter pulse is needed, a CD4528 monostable, or a LM393 configured as a 'one-shot' monostable, can be inserted between the output of the CD4093 and the amplitude potentiometer. This would allow a variable length pulse to be generated at a fixed frequency. A circuit for the LM393 is shown in **Fig 8.92**.

The duration of the pulse is determined by the resistor connected to ground on the non-inverting input (R) and the capacitor (C) connected from here to the output. To make the pulse width variable, substitute a variable resistor.

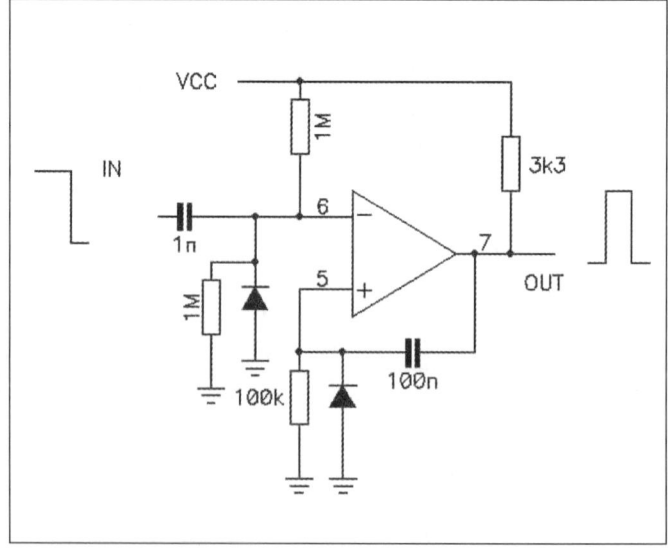

Fig 8.92 Optional monostable circuit using LM393

The formula to calculate the pulse period is:

T = CR

where R is in ohms and C is in Farads. For a resistor of 100kΩ and a capacitor of 100nF, the pulse width will be about l0ms.

The monostable is triggered by a falling edge voltage applied to the inverting input via a low value capacitor, typically 1nF. The potential divider across the inverting input needs to be made from two equal high-value resistors, eg 1MΩ. The two 1N4148 diodes prevent the inputs from being driven negative by more than 0.7V.

[1] *Power Supply Handbook*, John Fielding, ZS5JF, reprinted 2009 RSGB/ARRL (ISBN 9781 9050 8621 4)

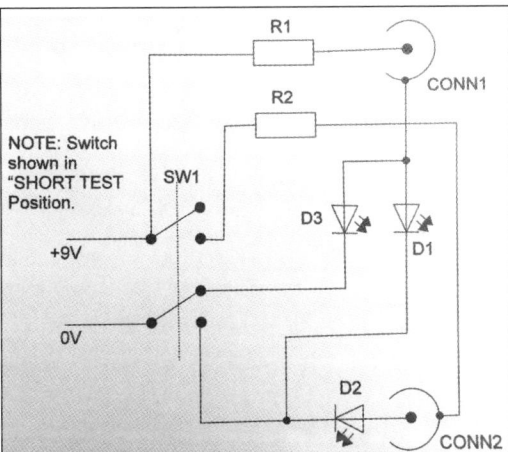

Fig 8.93 Coaxial cable checker

R1, R2	680R, 0.25W
D1, D2	LED, green
D3	LED, Red
SW1	DPDT
CONN1, CONN2	to suit

Table 8.21 Component list for the coaxial cable checker

8.16 Coaxial Cable Tester

As an aid to quickly performing DC checks, for example on coaxial cables to which connectors have just been fitted, Clive GM4FZH has designed a coaxial cable tester. This is a handy test-box into which a coaxial cable can

be plugged for a visual confirmation of DC continuity and the detection of any short-circuit (**Fig 8.93**). A components list is given in Table 8.21.

Construction Notes:

- The tester MUST be built into a plastic box

- Other types of connector can be wired in parallel with CONN1 and CONN2 to enable testing of a wider variety of cable terminations

- LEDs can be mounted in holders to ease build

- Designed to work from a PP3 battery or other 9V source

- Resistors R1 and R2 can be recalculated for other supply voltages

- An additional switch can be placed in the +9V line to provide an ON/OFF function

Method of Operation:

- Plug the cable to be tested between CONN1 and CONN2

- With the switch in the position shown, if there is a short in the cable, the red LED D3 will be illuminated. This is the 'SHORT TEST'.

- With the switch in the other position, if the centre conductor is intact, D2 will illuminate green. If the screening (braid) is intact, D1 will also illuminate at the same time. If only one LED illuminates then there is a cable fault denoted by the LED which is not illuminated. This is the 'CONTINUITY TEST'.

In the absence of the above, and particularly when testing aerial feeders where it may not be practical to access both ends of the coaxial line at once, the DC checks can be performed manually:

(a) Make sure the far end is open circuit. Do an ohm-meter check between centre and braid, this should show an open-circuit. If not there is either a short between the braid and the centre conductor (it could be at either end), or physical damage to the cable.

(b) Short the far end of the coaxial cable. At the near end the ohm-meter should read a short-circuit. If not there is a break (a discontinuity) in either the braid or centre conductor or a faulty connector.

(c) If (b) reveals a discontinuity, and it is infact practical to bring both ends together, ensure that both ends are open circuit, and connect an ohm-meter between the braid ends. Observe the resistance, and repeat the measurement between the centre ends in order to determine which path contains the break.

Once the DC integrity of the cable has been verified, further checks can be made; for example:

Basic RF check

Connect a known good dummy load (eg 50Ω) at the far end. At the near end connect an antenna bridge, analyser, or RLB etc, and measure the impedance. Over a wide range of frequencies this should show a pure resistance of close to 50Ω. If not, check all connectors and look for cable damage (eg kinks, abrasions, cuts, discoloured braid).

Advanced checks

If one has access to an analyser it is also possible to check distance to fault, cable attenuation and so forth. See the analyser instruction manual for details.

Similar tests can also be made on a balanced feeder, taking the appropriate precautions and trying to keep it balanced with respect to ground.

Appendix A – Behind the Modern Multimeter

WITH THE ADDITION of a few components, a simple analogue panel meter, which might for example read 50uA full-scale, can be adapted to measure a wide range of DC current, voltage, and resistance. This forms the basis of what has become known as the analogue multimeter. With rectification, it can also measure AC voltage and current. Likewise, the digital panel meter, which is voltage rather than current driven (perhaps requiring 0.2V full-scale), can be adapted using similar techniques to make a digital multimeter. Described here is the theory and practice of how both types of panel meter work, and how they can be adapted to make a multimeter.

Consideration is also given to two particular types of meter driver, which enable a meter to be used as a high-impedance voltmeter or for measuring very low currents.

Section 1: Analogue Panel Meters

A1. Types of Analogue Meter

The two most common forms of analogue meter are the moving coil and the moving iron type. The former is the most common, has a linear response to current, and responds only to DC. In contrast, the latter tends to be found in power applications, has a non-linear response to the applied current, but responds to both AC and DC. Both types rely on a magnetic field produced by a current passing through a coil reacting with a permanent magnetic field, or a piece of iron respectively. The force so developed, drives a pointer which indicates a value on a scale. Although these meters are current-driven, voltage measurements can be inferred (see later). **Pic A1.1** shows a selection of analogue meters. The difference in construction and operation of these two types is explained below.

A1.1 Moving Coil

This type consists of a coil of wire wound on a former which is mounted on a spindle between pivots located within the magnetic field of a permanent magnet. Attached to the spindle is a pointer (balanced with a small counterweight at the other end) which passes across a scale. A typical arrangement is shown in **Fig A1.1**

External current is fed into the coil by means

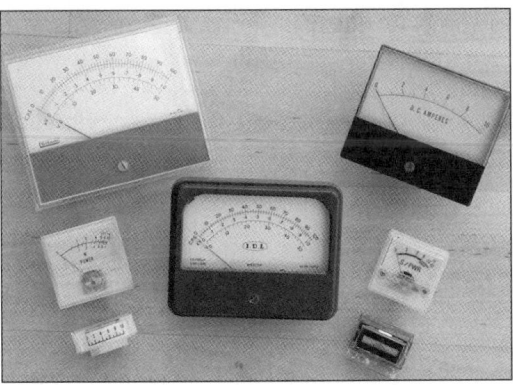

Pic A1.1 Various moving-coil and moving-iron meters

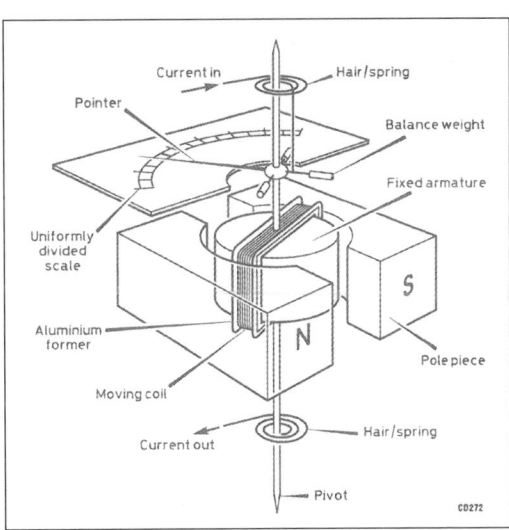

Fig A1.1 The moving coil meter

of two hair springs and produces an electro-magnetic field around the coil, which reacts with the magnetic field of the permanent magnet. The hair springs provide a load against which a resultant torque acts, the pointer deflection being proportional to the current passing through the coil. The hair springs also serve to return the pointer to a zero position when current flow ceases. In order to dampen the coil system, that is to make it 'less lively', it is usual to wind the coil on an aluminium former which acts as a short-circuited single-turn coil in which eddy currents oppose the movement.

The moving coil instrument will only respond to DC currents. The scale is linear and can be sub-divided with appropriate legends as required.

A meter is normally specified by the current required for Full Scale Deflection (FSD), the resistance of the coil, and the instrument's shape and size. The common range of moving coil meter movements is from 50µA to 10mA FSD, while the arc through which the pointer moves (and hence the scale length) varies between 90° to 270°, with 90° and 120° being the most common. In addition there are some meters which have a centre zero position.

A1.2 Moving Iron

This type of meter relies on the attraction or repulsion of a piece of ferromagnetic materi-al with the magnetic field produced by a fixed coil. **Fig A1.2** shows the basic construction of an *attraction* type.

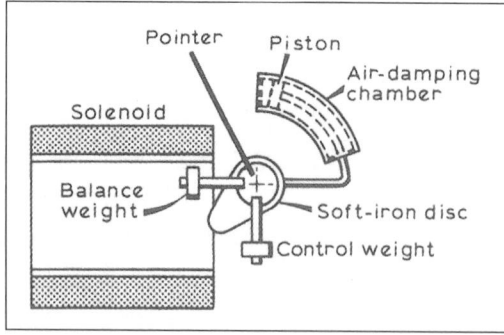

Fig A1.2 The attraction-type moving-iron meter

Here, a small piece of iron is drawn into the core of the coil when the current flows. It does not matter which way the current flows in the coil, magnetic attraction always takes place. In the *repulsion* type there are two pieces of iron rod within the coil - one fixed and the other movable and to which the pointer is attached. When a current passes through the coil both rods are magnetised in the same direction and since like poles repel, the moving part is repelled by the fixed part. It therefore does not matter which way the current flows in the coil, the result is always the same. Moving iron instruments therefore respond to both AC and DC.

In both types, control of movement is by hair springs, and damping is usually achieved by a vane in a cylinder containing air. The scale is non-linear, the graduations being very close at low values of current. At higher values of current, the graduations become more even, but are still basically non-linear. The typical meter movement starts at about 100mA FSD whilst the frequency range ex-tends to about 60Hz. Moving iron meters are not used for multimeters.

A2. Care & Use of Analogue Meters

A2.1 Storage and Transport

Apart from intrinsic internal damping, the de-gree of damping is also dependent upon the value of any external resistance placed in parallel with the coil. It follows that it is good practice to place a short circuit across the terminals of any sensitive instrument when not in use, or if being moved. As the mech-anism is delicate, any unnecessary vibration and mechanical shock should be avoided.

A2.2 Cleaning

Many older meters have their glass front glued onto the inside of the housing (PIC A2.1). If the glass must be cleaned, care should be taken not to let any cleaning fluid to run to the edges which may ingress and weaken the bond. Irrespective of this, min-

Pic A2.1 Care! Remember not to press too hard on the glass of old panel-mounted meters!

imal pressure should be put on the glass to prevent it collapsing inwards. If such an event should happen, it may be possible to re-glue the glass in place, but the glass may well have bent the very delicate pointer and perhaps damaged the hairsprings and coil, a catastrophe from which it may not be possible to recover. So be gentle!

A2.3 Mounting

Meters are normally balanced to work in stated orientations - horizontal, vertical, or at 45°. They may be used in other positions, but the zero may need to be adjusted and there may be small additional errors due to pivot friction. If mounted in a panel, the supporting material, such as ferromagnetic or aluminium sheet, may also affect the meter reading.

A2.4 Zero Adjustment

An adjustment screw, accessible from outside the instrument, allows a small twist of the hair springs so that the pointer can be accurately zeroed. As it connects with the springs it may be electrically live if made from a conducting material – so beware!

A2.5 Meter Sensitivity

The sensitivity of a voltmeter is usually expressed in Ohms/Volt (Ω/V). This is merely the reciprocal of the full-scale current sensitivity, I_{FSD}, of the basic meter. Hence, a 1mA meter used as a voltmeter would be described as 1000Ω/V and a 50µA meter as 20,000Ω/V.

A2.6 Meter Resistance

A good quality meter will have its I_{FSD} and coil resistance (which is essentially the same as the meter resistance) marked on it somewhere, often on the scale plate. The usual range for different movements is:

50µA	1800-3000Ω
100µA	1200-1800Ω
1mA	70-200Ω
10mA	5-15Ω

However, when visiting, for example, radio rallies or junk sales, one may pick up meters with only the I_{FSD} value given, so the question then is how to obtain the meter resistance, without overloading and damaging the meter, which is likely if a commercial ohmmeter is simply connected across it. To overcome this problem, the value can be determined by an indirect method, as shown in **Fig A2.1**.

Connect the circuit, choosing a value of resistance R1 and voltage V to give a reasonable deflection on the unknown meter. With S1 open, adjust R1 to give full scale deflection. Close S1 and then adjust R2 to give half scale deflection on the meter. Disconnect the voltage source, remove R2 and measure it

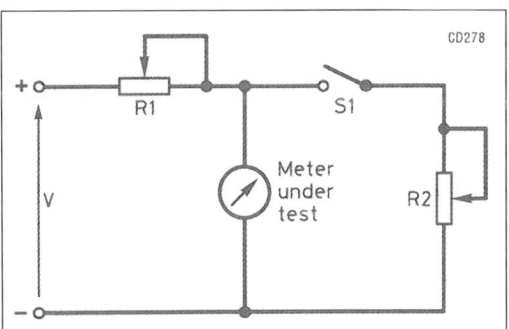

Fig A2.1 Circuit for determining meter coil resistance

with an ohmmeter, this will be the value of the meter resistance.

A2.7 Meter Protection

Meters are relatively expensive and are easily damaged if subjected to excessive current. Not only is the coil likely to burn-out, but damage to the pointer and other components of the mechanical movement is also possible. Clearly, it is advisable to provide some form of electrical protection provided it does not significantly degrade its operation.

A2.7.1 Parallel Diodes

Damage can be prevented simply and cheaply by connecting two silicon diodes in anti-parallel (anode to cathode) across the meter terminals (**Fig A2.2**) and this should be regarded as standard practice as no perceptible change in sensitivity or scale-shape need occur.

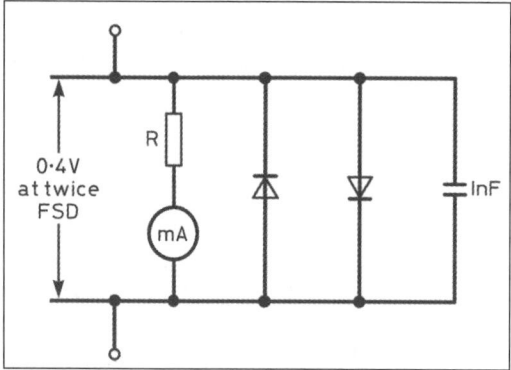

Fig A2.2 Meter protection using diodes

Theory: A characteristic of silicon diodes is that they remain very high resistance until the anode is some 400mV above the cathode at which point they start to conduct and the resistance falls to a low value. Since the voltage drop across the average meter is around 200mV, it follows that a silicon diode connected across the meter will have no effect even when the meter shows full-scale deflection. If however, the meter is overloaded to twice FSD and the voltage across the

meter rises to 400mV, the diodes will begin to conduct and shunt the meter against a further increase in fault current.

A2.7.1.1 Choice of Protection Diodes

Often, small silicon signal diodes such as the OA202, 1N914 or 1N4148 will be satisfactory. These have an inherently high reverse-resistance, or in other words a low reverse-leakage current, which minimises their shunt effect.

However, it is important that under the worst fault conditions the diode will not fail, going open circuit and affording no protection. An example of this would be in a high voltage supply where a large current could flow in small signal diodes in the event of a short circuit of the power supply. Here, a rectifier diode could be used instead, such as the 1N400X or 1N540X series. The reverse current of these diodes may be a few µA and, depending on the current to be measured, may have a slight effect on the sensitivity of the meter circuit.

Zener diodes can also be used for meter protection, particularly those with avalanche characteristics (ie a sharp knee), which applies to diodes of 6.2V or more. However, there is rarely any advantage in the use of Zener diodes, and in fact there could be a disadvantage as they usually show notable reverse leakage. They are useful for meters which will not stand an overload significantly greater than their FSD, or which have a voltage drop larger than the turn on voltage of a silicon diode. If Zener diodes are used, they should be connected in series as shown in **Fig A2.3**.

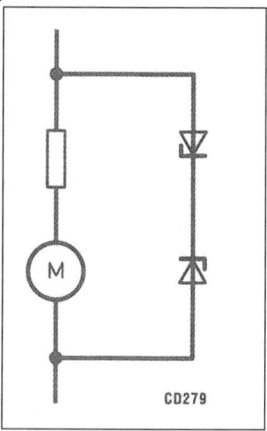

Fig A2.3 Use of Zener diodes for meter protection

A2.7.2 Adverse Effects in Presence of AC

Although diode protection should be applied as routine in order to safeguard instruments, it can cause some unusual effects if measurements are made with an AC signal imposed on a DC signal. This AC component, providing it is symmetrical, should not introduce any error. However, if the AC were large enough, it may bring the diodes into conduction at the peak of the cycle, thus introducing a dynamic shunt on the meter. This can be confusing when back to back diodes are used, as the meter sensitivity will drop without any offset reading to warn what is happening. These effects are most likely to occur when measuring rectified mains or when RF is present.

A2.7.3 Series Resistance

It is wise to include a series resistor as shown earlier in **Fig A2.2** such that the voltage drop across the meter/resistor combination is 200mV at FSD.

A2.7.4 Parallel Capacitance

Whenever a meter is to be used and RF may be present, it is wise to shunt the meter with a capacitor, typically a 1000pF ceramic type, as shown in **Fig A2.2**. If strong RF fields will be present, for example in a transmitter, it would also be wise to shield the meter and possibly feed it via shielded cable. These protection methods should culminate in the scheme shown in **Fig A2.4**.

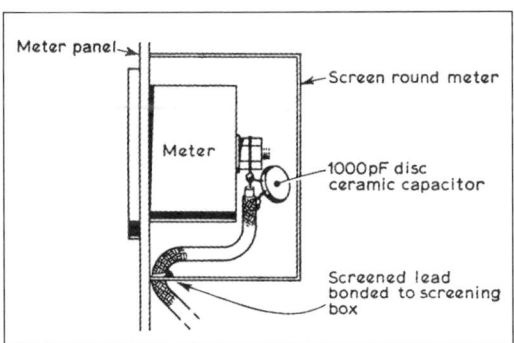

Fig A2.4 Screening and bypassing a meter in a transmitter

A2.8 Extending the Current Range – 'Shunts'

Meters commonly have a Full Scale Deflection in the range 0-3 or 0-10, or multiples thereof, which may not be convenient for the measurement required. However, by the addition of an external 'shunt' resistance it is possible to extend the range of the meter while still using the original scaling. A typical circuit is shown in **Fig A2.5**.

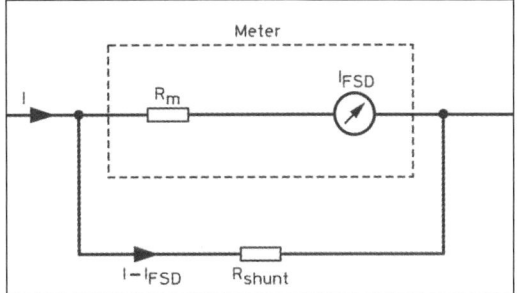

Fig A2.5 Adding a current shunt

Theory: This technique diverts a proportion of the current to be measured away from the meter so the the FSD current is not exceeded. Here, assuming the maximum current to be measured is I, then the shunt resistance and its maximum dissipation is given by:

$$R_{shunt} = R_m * I_{FSD}/(I - I_{FSD})$$
$$P_{shunt} = R_{shunt} * (I - I_{FSD})^2$$

where I_{FSD} is the current at full scale deflection of the meter, and R_m its resistance.

It is generally convenient to choose I to be a multiple, or sub-multiple, of the maximum scale reading so that the scale reading is easily multiplied. The multiplying factor is given by:

$$n = 1 + (R_m/R_{shunt})$$

If desired, and if practicable for the meter in question, the scale could be redrawn to reflect the new range, perhaps using transfers. Alternatively, there is software available for such a task – see the beginning of Chapter 8 for links.

A2.8.1 Making Meter Shunts

Shunts can be found for sale on Internet auction sites, but they tend to be for specific meters such as the (now aged) AVO8; however It is possible to make shunts. For values of 0.1Ω or greater, series/parallel combinations of standard off-the-shelf resistors may be used. Otherwise they can be made of resistance wire, or standard enamelled copper wire.

Appendix C provides a wire table which gives the equivalent diameter of wire in imperial and metric units. The current carrying capacity and the resistance per unit length can be found on the Internet or from a catalogue. Compute the resistance required, choose a wire capable of carrying at least the current required and calculate the length required. This should then be wound on a former such as high resistance 1W or 2W carbon resistor. Always use as thick a cable as possible in order to minimise the heat rise in the shunt due to the I^2R loss. Ideally, obtain constantan (nickel and copper) or nichrome (nickel and chromium) resistance wire, which has a very low temperature coefficient.

By way of example, consider a shunt designed to convert a 10A FSD meter to 20A. The meter develops 75mV across it when 10A flows through it. What length of resistance wire at 0.25Ω/m will be required?

With 10A flowing through the meter, 10A will have to be bypassed though the shunt wire, therefore:

$R_{shunt} = 0.075V/10A = 0.0075\Omega$

Length of resistance wire required = 0.0075/0.25 = 30mm.

Note: it would be almost impossible to make this out of a series of parallel resistors. Even with resistance wire, great care must be exercised in making the end connections to have minimal resistance. The shunt could probably be made from a piece of bar, providing its resistivity is known.

A2.9 Voltage Measurement

To use a meter as a voltmeter, the maximum voltage to be read provides the value of I_{FSD}.

The circuit used in this case is as shown in **Fig A2.6**. The value of the series 'multiplier' resistor R_{mult} and its maximum dissipation P_{mult} is determined by:

$R_{mult} = (V/I_{FSD}) - R_m$

$P_{mult} = R_{mult} * (I_{FSD})^2$

For example; a 50μA meter with a coil resistance of 3000Ω is required to measure up to 30V. What value of multiplier resistor is required, and what is the thermal dissipation in this resistor? Answer: $597k\Omega$ and 1.5mW.

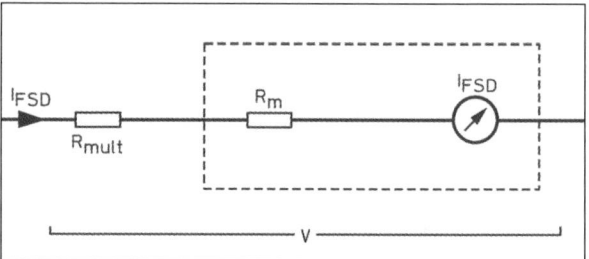

Fig A2.6 Adding a voltage multiplier

A2.10 Measurement Errors

It is imperative to remember that the meter will load the circuit to be measured. The key is to recognise what can make the measured value incorrect and how to minimize the impact of the test instrument.

A2.10.1 Voltmeters

For example, putting a meter which requires 50μA, across a resistor through which only 100μA flows, will disturb the circuit significantly. Putting the same meter across a resistor through which 10mA flows will have little effect.

How can one gauge this or guard against it? Consider a $20,000\Omega$/V meter, set on the 10V range: this will have a resistance of 10 x 20,000 = $200k\Omega$. It is suggested that any resistance across which this voltmeter is placed should have a maximum value of one tenth of this, ie $20k\Omega$. For any range, one can use this rule of thumb method. Put another way, the effective meter resistance should be at least 10X the resistance to be measured. The greater the ratio, the more accurate will be the measurement.

A2.10.2 Ammeters

For ammeters, the voltage drop across the ammeter (=$I_{FSD}*R_m$) is significant in relatively low voltage circuits. For example, a 0.5V drop across an ammeter is unacceptable in a 12V circuit but immaterial in a 100V circuit. The trick is to choose a voltmeter that has as low an R_m as possible. This reduces the in-circuit voltage drop and keeps any shunt resistance as high as possible. If possible, use an ammeter of I_{FSD} equal to, or just greater, than the range required.

A2.10 Choice of Meter

From the foregoing it can be appreciated that the correct choice of meter requires needs an understanding of the circuit to be tested. For typical DC circuits used by Radio Amateurs, a moving coil meter will be needed in the vast majority of applications.

For voltage measurements the meter should take only a small current compared to what flows in the circuit. Make the current drawn by the voltmeter circuit less than 5% of what can be supplied. The smaller the percentage, the more accurate will be the reading.

For current measurements the voltage drop across the ammeter must be kept as low as possible. Choose a meter that has as low a coil resistance as possible and an I_{FSD} equal to, or just greater than the range required. Try to keep the voltage drop less than 5% of the supply voltage. **Fig A2.7**

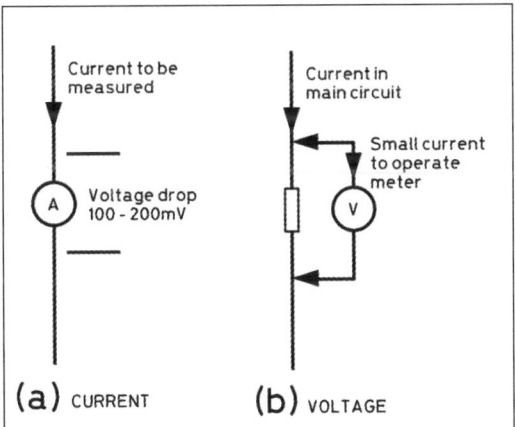

(a) CURRENT   **(b)** VOLTAGE

Fig A2.7 Use of meters for measurement

For mains circuits of 100V or above, the moving iron meter may represent a more viable alternative and tends to be cheaper.

A3. Embedded Meters for Switching (Ranging)

To save cost (and sometimes panel space), a meter may serve several functions, especially in for example, valve circuits where it may be measuring grid and anode voltages and grid and anode currents, these normally require different ranges for the various parameters being measured. For convenience, two meters would be used - a voltmeter and an ammeter. In all instances, a break before make switch should be used; particular care should be exercised in selecting the switch when high voltage circuits are involved.

It is important that the resistance of wire and switch contacts do not become significant compared to the value of low resistance shunts, or (as we have discussed) meter readings will become erroneous. By contrast, the switching-in of different multiplier resistors is of little consequence as these tend to be high value resistances.

A3.1 Voltmeters

For a switched embedded meter, voltages are often measured with respect to 0V, which means that one side of the voltmeter usually connected or 'tied' to chassis ground. On the other side, the various series resistances can be calculated from the characteristics of the meter. It is suggested that the lowest val-

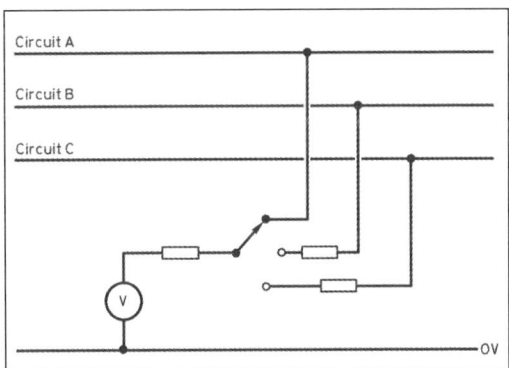

Fig A3.1 Switched voltage measurements

ue is wired directly in series with the meter, and other values chosen such that this value, plus the additional one, equals the value calculated. This is shown in **Fig A3.1**, where A is the lowest voltage to be measured. This ensures that some current limiting always exists in series with the meter giving it a measure of protection.

A3.2 Ammeters

For the purposes of this section, a meter of 1mA FSD and a coil resistance of 100Ω is assumed. **Fig A3.2** shows how switching could be arranged for the measurement of current in three ranges. Switch/conductor resistance is unlikely to be a problem with circuit A, it may be a problem on circuit B and certainly will be on circuit C.

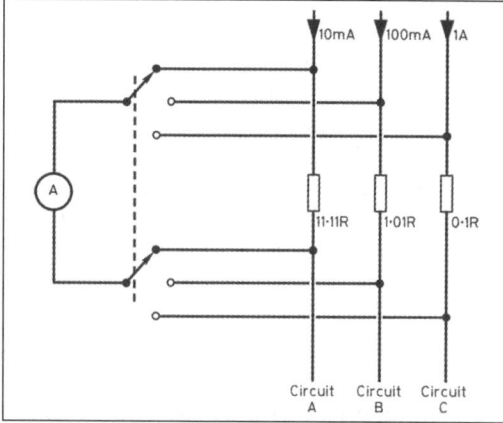

Fig A3.2 Switched current measurements

One solution is to use a non-switchable meter for any current range which requires a low shunt value, typically less than about 0.5Ω. A different approach is to use the meter to measure the volt-drop across a resistor. If a

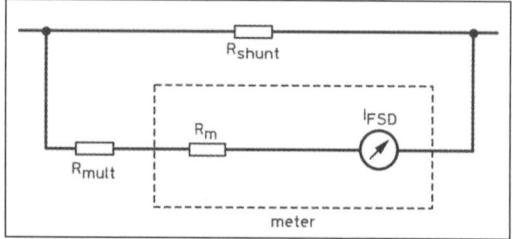

Fig A3.3 Current measurement by volt-drop method

50μA meter with 3kΩ coil resistance were to be used, then a minimum voltage drop across the meter of about 150mV is required, while for a 1mA movement with 100Ω resistance, it falls to about 100mV. The voltage to be developed should be equal to or greater than $I_{FSD}*R_m$. The circuit used for this arrangement is given in **Fig A3.3**.

A4. Making AC Measurements

A4.1 Moving Coil Meters

If an alternating current is passed through a moving coil meter there will normally be no deflection since the meter will indicate the mean value, and in the case of a waveform symmetrical about zero, this is also zero. If however the AC is rectified, such that the meter sees a series of half-sine pulses (full-wave rectification), the meter will indicate the mean value ($2/\pi$ or 0.637 of the peak value). Commercial instruments using moving-coil instruments for AC sinewave measurements therefore incorporate a rectifier arrangement (see **Fig A4.1**) and the scale is adjusted to read RMS values (0.707 of the peak value). *They will read incorrectly on any waveform that does not have these relationships.* The moral is, do not use a moving coil meter on any waveform other than a sinewave. This rectifier arrangement is normally only used for voltage measurements; AC current measurements pose additional problems and are not considered further.

A4.2 Moving Iron Meters

Moving iron instruments, as previously men-

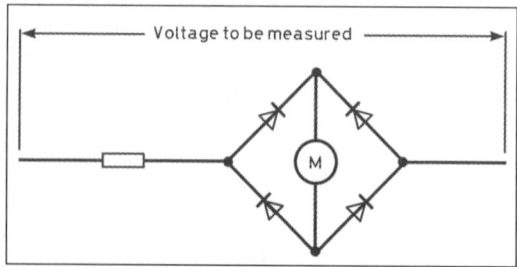

Fig A4.1 Typical arrangement for AC voltage measurements

tioned, do respond to an AC current and can be used for measurements without rectifiers. This type of meter unfortunately has a square law characteristic and so the scale tends to be cramped at the lower end. They normally have a full scale reading of about 20% more than the normal value to be displayed. As noted earlier, moving iron meters are not used for multimeters.

A5. Panel Meter Drivers

This section deals with instruments which have active circuits preceding a pointer-type instrument to increase its sensitivity, input impedance, and frequency range. Not all of these may be achieved with one instrument - it depends what it is has been designed for.

With the following types of circuits, full scale deflection can be achieved with only a few millivolts or micro-amps. Input impedances up to about 50MΩ can be attained, which reduces the loading on a circuit under test. In conjunction with diode probes, accurate AC measurements can be made in excess of 100MHz.

A5.1 Hi-Impedance Adapter for DC Voltmeter Measurements

The typical analogue electronic meter uses operational amplifiers to provide high input impedance, buffering and amplification if necessary. It is quite common for the op-amp to drive a meter with 1mA movement. How can this be achieved? **Fig A5.1** shows a circuit for measuring DC voltages with a 10V and 30V range. The TL071 op-amp has very high input impedance and is used in a unity, non-inverting gain mode. On the 10V range the input resistance is R2+R3; on the 30V range it is R1+R2+R3. These resistors also act as attenuators. The arrangement with R2 always in series with the input is preferred as it provides some protection for the op-amp. RV1 and R4 provide offset balance null due

to the op-amp providing a small DC bias at the output. Resistor R5 limits the current through the meter and can be varied to suit various meter sensitivities; it may be better split into two components in the ratio of about 75% fixed and 25% variable. The use of a reasonable output voltage and a 1mA meter means that the effect of the meter resistance is almost negligible. For a 1mA meter movement, it is suggested that R5 is about 8k2. Diodes D1 and D2 are used to protect the meter movement as explained earlier.

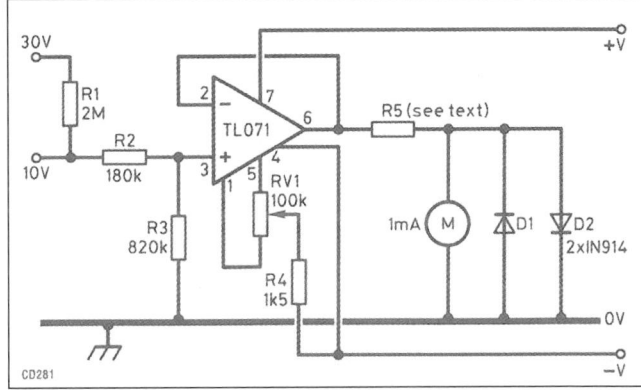

Fig A5.1 Basic high-impedance DC voltmeter

A5.2 Very Low Current Adapters for DC Measurements

To measure DC current it is often necessary to measure the voltage drop across a resistor through which the current flows. As discussed earlier, the amount of voltage drop that can be tolerated is less than 200mV in order to minimise any disturbance to the circuit being measured. For example, **Fig A5.2** shows 1mA flowing through 100R, producing 0.1v. This voltage meets the criteria, but must be amplified to provide a useful indication on the meter. A suitable non-inverting amplifier is shown, the gain of which is given by (R1+R2)/R2 and with the values given, comes to approximately 82 (ie 8.2V/0.1V). The value of R3 for a typical 1mA movement should be approximately 8k2, possibly a 6k8 in series with a 2k2 trimmer.

Developing this idea, the circuit in **Fig A5.3** and component list in **Table A5.1** will measure

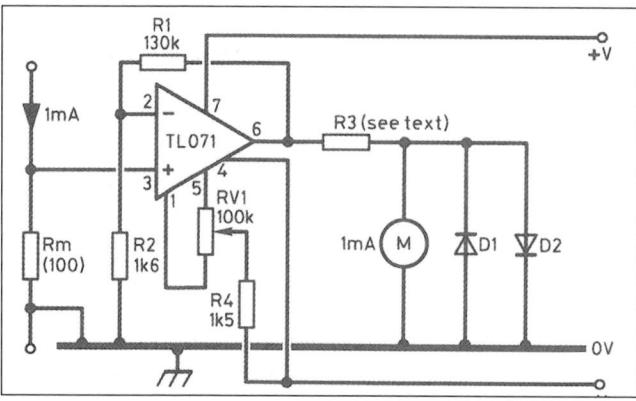

Fig A5.2 Basic arrangement for current measurement

current from under 1µA up to 10mA in five ranges. It uses the principles outlined earlier by converting the current to a voltage and using an operational amplifier to amplify the signal and drive a meter. The advantage of this arrangement is that the voltage dropped across the resistor is low - in this case 10mV maximum on any range. If the circuit is battery powered, then it can be used floating; with a split mains supply the current is usually referenced to ground with the arrangement suggested.

Because of the resistance values required in the input, it is necessary to construct them from two parallel units. The op-amp used should have a very low input bias current - the TL071 is quoted as 30pA.

Construct the circuit as shown and apply a split power supply. No special precautions are required. A PCB layout is given in Appen-

dix D. With no input, adjust trimmer RV1 to obtain zero deflection on the meter. The unit is now ready for use.

A5.3 High Impedance AC Voltmeter Measurements

The circuits discussed so far have been for DC only. The problem for AC circuits is to convert the signal to DC as soon as practicable and then measure it. The other problem is the voltage drop across the rectifying diodes. A solution can

R1a,1b	18k, 0.5W, 1%
R2a,2b	1k8, 0.5W, 1%
R3a,3b	180R, 0.5W, 1%
R4a,4b	18R, 0.5W, 1%
R5	1R, 0.5W, 1%
R6	10k, 0.5W, 1%
R7	1k, 0.5W, 1%
R8	1k5, 0.5W, 5%
R9a	100k, 0.5W, 1%
R9b	10M, 0.6W, 1%
R10	1k, 0.5W, 5%
RV1	100k skeleton pot.
IC1	TL071
SW1	Switch, 1p, 5w
M1	1mA FSD meter

Table A5.1 Component list for the DC micro/milli-ammeter

be achieved with a precision rectifier which consists of op-amps with diodes in the feedback loop in order to minimise the effect of the diode forward voltage drop.

The meter presented here is for analogue signals and is reasonably accurate from 10Hz up to about 0.5MHz. It has four ranges, corresponding to 10mV, 100mV, 1V and 10V, the output meter being switchable to read either RMS or peak for a sinewave input. The input impedance is 1MΩ shunted by a few pF, which depends on the method of wiring, sockets used etc.

Fig A5.3 A DC micro/milli-ammeter

R1,2	1M1, 0.6W, 1%
R3	4M3, 0.6W, 1%
R4	5M6, 0.6W, 1%
R5,7,8	1k, 0.6W, 1%
R10-15	1k, 0.6W, 1%
R6,9	10k, 0.6W, 1%
R16	4k7, 0.5W, 5%
R17	10k, 0.5W, 5%
RV1,2	5k trimmer
C1	1µF, 50V
C2,3	100µF, 25V
C4,5	100nF, 50V ceramic
C6	10µF, 6.3V
D1-4	1N914 or similar
IC1	TL071
IC2,3	NE5532
S1	3p, 4W rotary switch
S2	SPCO switch
M1	Meter, 100µA FSD

Table A5.2 Component list for high input impedance voltmeter

Circuit description

The circuit diagram is given in **Fig A5.4** and the component list in **Table A5.2**. The input is buffered by IC1, a FET input op-amp; R1 to R4 form a constant input resistance network providing an attenuation of 10 on the 10V range. Amplifiers IC2a and IC2b each have a gain of 10, the 10mV and 100mV ranges being routed via these. Switch SW1 selects the correct amplification/attenuation for the ranges such that a 1V signal is presented to the precision rectifiers formed by IC3a/b. The DC output from these rectifiers is then scaled for peak or RMS by networks R16/RV1 and R17/RV2, the resulting output driving meter M1. IC2 and IC3 are formed by dual op-amps type NE5532 which has a 10MHz unity-gain bandwidth product.

Construction and calibration

Providing the components listed are used there should be little problem in constructing the circuit and getting it to work. To assist, a PCB layout pattern is given in Appendix D. The unit should be well screened in a metal box to avoid extraneous pick-up. To calibrate, a known amplitude sinewave should be applied to the input, the correct range selected, and RV1 (peak) and RV2 (RMS) adjusted for the correct reading.

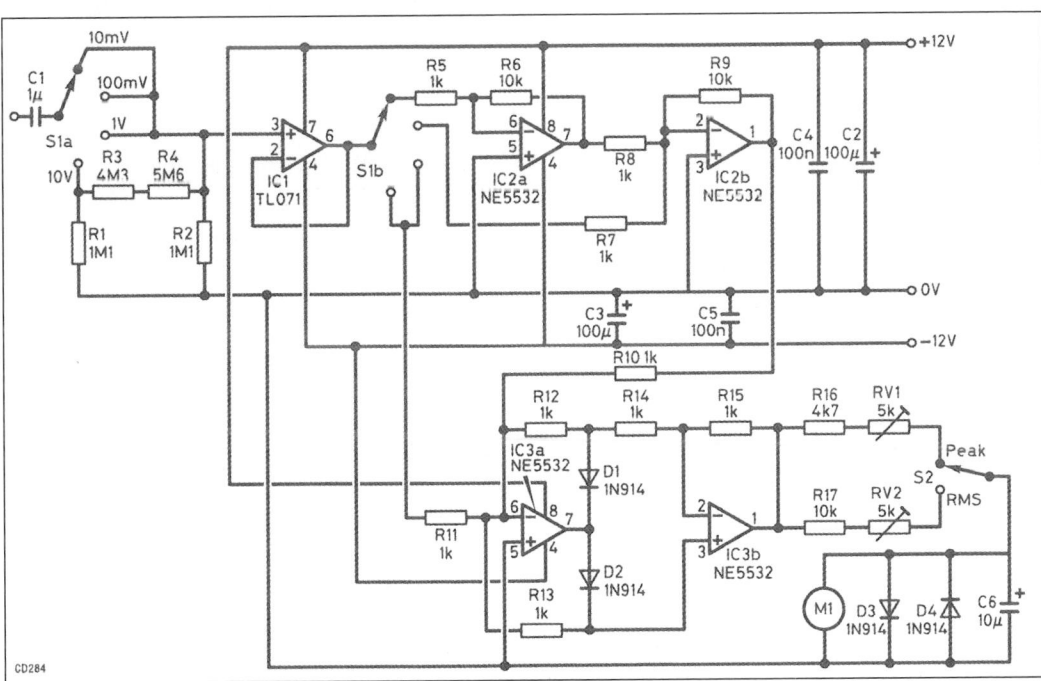

Fig A5.4 High input impedance AC voltmeter

Section 2: Digital Panel Meters

The digital meter has become a relatively cheap and ubiquitous item, and is often less expensive than its analogue counterpart. Although it can only jump in discrete steps (hence 'digital'), and requires a power supply, it is capable of great accuracy and provides an easy readout of the measured value. In contract to analogue panel meters, the digital meter is voltage, rather than current driven.

A6. Theory of Operation

The digital meter works by converting an input analogue voltage (or a current converted to a voltage) into a digital signal that can be used to drive either an LED (Light Emitting Diode) or LCD (Liquid Crystal Display) - see **Fig A6.1**. The conversion technique used is either a straightforward A to D converter or the dual ramp technique. A digital voltmeter is often quoted as having, for example, a 3½ digit display. This means that it will display three digits (0-9) with the most significant being only a 0 (normally suppressed) or a 1, ie a maximum display of 1999.

There are quite a few ICs made by various manufacturer's that provide a basic digital voltmeter, with external components being required for extending the range, over voltage protection, and display functions. These

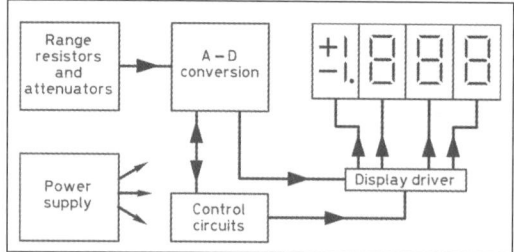

Fig A6.1 Block diagram of digital meter

ICs provide outputs suitable for driving LEDs, LCDs, or BCD processors etc.

A7. Digital Panel Meter Modules

These include the above ICs and come with a 3½ or 4½ digit display and are capable of

accepting a 200mV signal. These are relatively cheap and provide a good basis for making various types of metering system.

Based on an LCD, they either plug into a DIL socket, or are on a small PCB - see **Fig A7.1**. They usually read full scale for a 199.9mV DC input, work from 5V or 9V DC supplies (depending on model), consume very low current (typically 150 - 300μA on a 9V supply) and have an input resistance of at least 100MΩ. The panel meter will provide an accuracy of 0.1% or better, although this does not take into account any external signal conditioning circuits such as amplifiers or attenuators. In addition to these parameters, displays often show units such as μ, m, V, A, Ω, Hz etc (referred to as annunciators). They can be purchased with and without back-lighting.

The main design consideration in using these units is to get the parameter to be measured into a DC voltage in the range 0-199.9mV. Circuits for doing this can include amplifiers, attenuators and rectifiers.

Fig A7.1 Photograph of typical digital panel meter

A8. Meter Interfacing to an External Voltage or Current

In the same way that resistive shunts or multipliers are used to condition the current into an analogue meter, so a similar arrangement is needed to condition the voltage into the digital meter. Measures such as diode protection may be similarly applied.

A8.1 Voltage Multiplier

Fig A8.1 (component list **Table A8.1**) shows

R1	10M, 0.6W, 1%
R2	1M, 0.5W, 1%
R3	100k, 0.5W, 1%
R4	11k, 0.5W, 1%
R5	110R, 0.5W, 1%
R6	100k, 0.6W, 5%
D1,2	1N914 or similar
PM1	Anders Panel Meter type OEM22
SW1	Rotary switch 3p, 4w

Table A8.1 Component list for the digital voltmeter

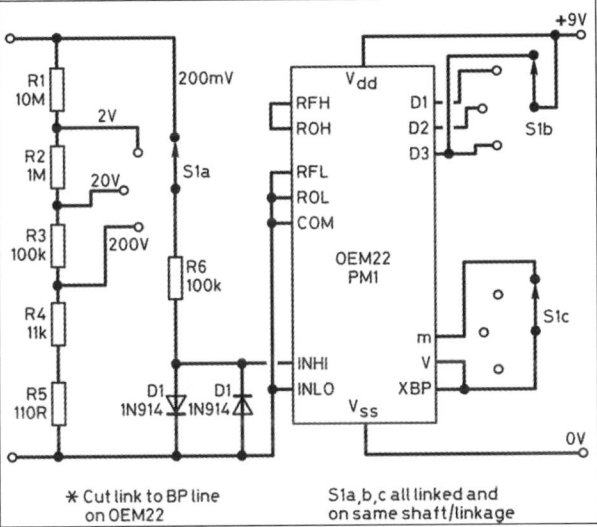

Fig A8.1 A practical digital voltmeter

the arrangement for a digital voltmeter with DC ranges of 200mV, 2V, 20 V and 200V. Resistors R1 to R5 form a potential divider network with switch SW1a selecting the correct input - ie a maximum voltage to the panel meter of 199.9mV. Resistor R6 and diodes D1/D2 provide protection for the panel meter should the wrong range be inadvertently selected. For inputs lower than 200mV an amplifier is required ahead of the meter input.

A8.2 Current Shunt

As we have seen, the voltage drop to be developed by the current passing

through the measurement resistor must be 200mV for full scale. Hence for a digital meter to measure 200μA to 200mA in decade ranges, the circuit in **Fig A8.2** results. Resistors R1 to R4 form the load across which the voltage is developed from the current being measured; resistors R1 to R3 involve resistors in parallel to make up the correct value. Switch SW1a selects the input. The combination R5/D1/D2 provides protection for the panel meter input.

A9. Level Indicators

Another method of indicating the value of a parameter is to use solid state indicators. These consist of a row of LEDs which illuminate when a certain value has been exceeded, the differing thresholds being set by the driver IC. The indicators can be separate LEDs, or a bar indicator consisting of a row of LEDs with perhaps a common cathode or anode. Indicators come in a variety of forms and may consist of typically 2, 3, 4, 5, 10, 30, and 101 elements. They can run in two modes; the dot-mode where an LED illuminates indicating that value; or the bar-mode, where all LEDs below that value also illuminate. This function is selectable on the driving IC. In the dot

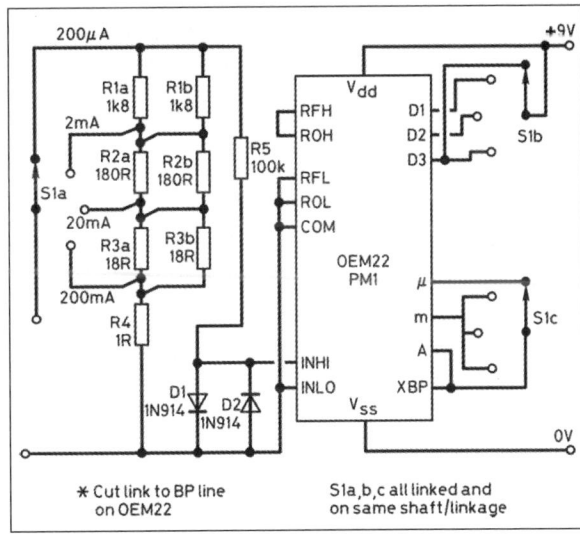

Fig A8.2 A practical digital ammeter

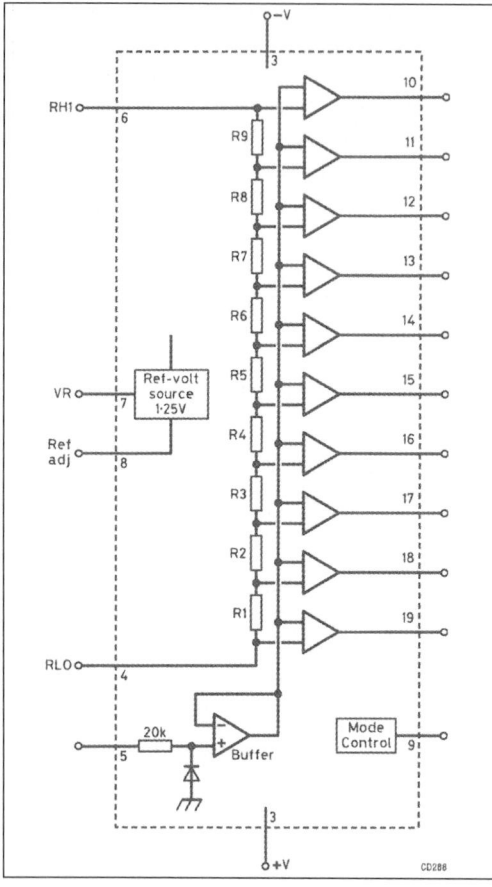

Fig A9.1 Typical bar-graph driver IC

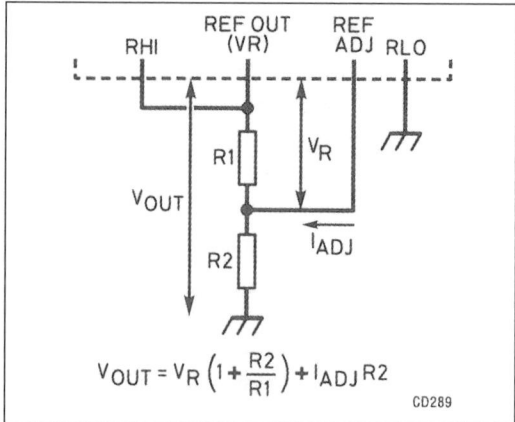

$$V_{OUT} = V_R \left(1 + \frac{R2}{R1}\right) + I_{ADJ} R2$$

CD289

Fig A9.2 Modification to reference output

not less than V- and 1.5V below V+. Mode select gives either dot or bar displays.

The voltage reference source need not be used or it can be modified as shown in **Fig A9.2**. With VR being connected to RHI, and Ref Adj being connected to 0V along with RLO, the reference voltage (1.25V) is divided along the R1 to R10 chain; in the linear model this represents increments of 125mV. If it is modified, the equation of **Fig A9.2** holds, the current through R1 being given by R1/VR and should lie between 80µA and 5mA. Note also that the current drawn out of the reference pin determines the LED current, the LED current being about 10 times this current. This current should include both the current through R1 and R2 plus that through the divider chain. An external voltage can

mode, there is a slight overlap, so that at no point are all the LEDs extinguished.

There are three common driver ICs for these displays, each controlling ten LEDs. These ICs are:

LM3914 control in linear steps

LM3915 control in 3dB steps (logarithmic)

LM3916 VU Scale

Using the LM3914 as an example, a block diagram of this is shown in **Fig A9.1**. The resistors R1 to R10 set the various threshold levels for the comparators that drive the LEDs. The input signal is buffered, and then applied to the comparators. V- is normally ground (0v) and the maximum supply voltage (V+) is 25V. The input signal is restricted to a range

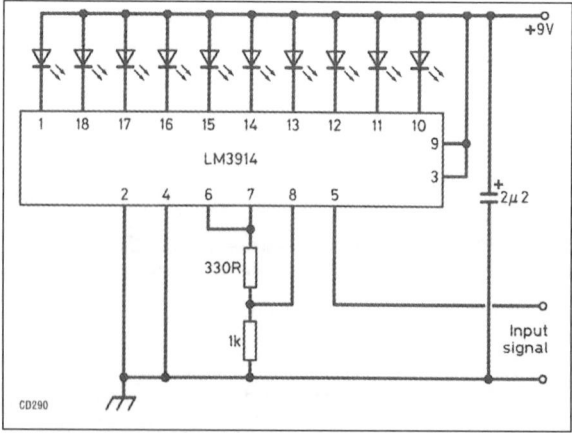

Fig A9.3 0-5V level detector

also be used to set the divider reference. In bar mode, current consumption for the LEDs should be kept to a minimum because of total IC dissipation - so keep the LED current to about 10mA. For further details see the IC data sheet.

A9.1 A 5v Level Detector

This circuit (Fig A9.3) is suitable for monitoring signals in 0.5V steps up to 5V. It can cope with DC levels and slowly varying AC signals. The unit is in bar-mode display with pin 9 tied to +V. For a dot display, open-circuit pin 9. The input signal is applied via pin 5 and ground.

Precautions required in using the circuit, are adequate decoupling; keeping leads to the LEDs short; and a common grounding point. It should be possible to modify this circuit for other ranges and add a diode detector for AC signals.

Section: 3 Creating a Multimeter

If a selector switch is used to apply the appropriate shunt or multiplier described earlier, a single meter can be used to measure voltage and current over multiple ranges, giving the arrangement the nickname 'a multimeter'. If an internal battery is included, resistance can be measured as well, by application of Ohm's Law.

A.10 Analogue Resistance Measurement

Two common arrangements are shown in Fig A10.1. Circuit (a) simply has a battery in series with a meter, a resistor RV1 and the unknown X. Circuit (b) has a refinement in the form of RV2 which can adjust the sensitivity of the meter to compensate for any drop in battery voltage. The latter circuit gives rise to the 'zero-ohm' knob which is fitted as standard on commercial analogue multimeters.

To use and calibrate, the terminals across which X will be connected are shorted together and RV1 adjusted until the meter reads full scale. When an unknown X is connected across the terminals the meter will read less than full scale. Calibration of the (non-linear) meter scale should be carried out using a series of known value, high-accuracy resistors. Alternatively a graph relating meter current to resistance can be prepared.

Fig A10.2 shows a multi-range instrument for DC voltage and current. It is based on a 50µA movement of resistance 2700Ω.

A.11 Digital Resistance Measurement

A digital panel meter is used in a circuit in which a known current passes through an

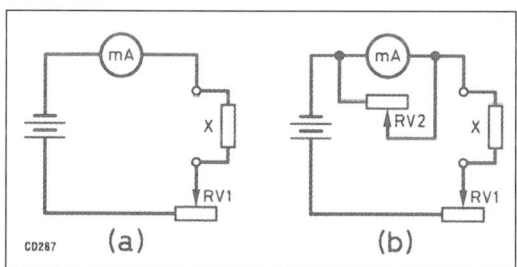

CD267 (a)   (b)

Fig A10.1 Two common arrangements for measuring resistance

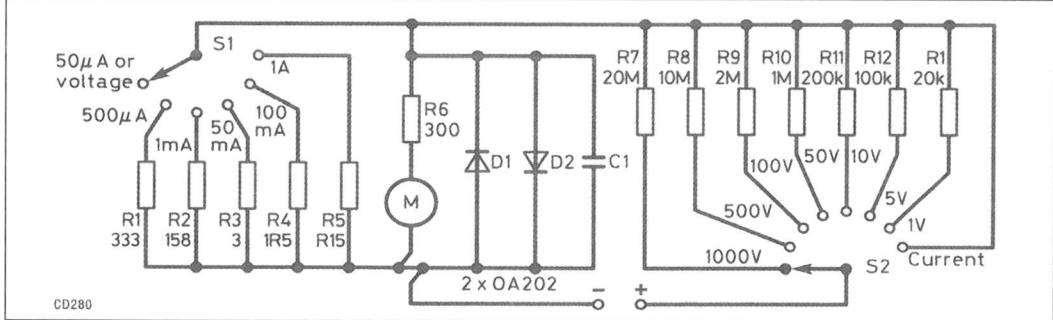

CD280

Fig A10.2 Basic analogue DC multimeter: current to 1A, voltage to 1000V

unknown resistor and the voltage across it is measured. Using Ohm's Law (V=IR) and a fixed value of I, causes V becomes a measure of the resistance. If I is made a multiple of 10, the display will indicate the resistance up to the scale display of 1999. All that needs to be fixed is the decimal point and the multiplier.

Where a digital rather than analogue meter is used, it will normally be designed to have the ability to self-zero and the opportunity will often be taken to include yet more features, such as the measurement of inductance, capacitance, and temperature, and to perhaps incorporate some limited frequency measurement and signal generation capability. Such instruments are now a relatively cheap and an indispensable item of great versatility representing excellent value for money.

A.12 Measurement Accuracy

A typical cheap analogue meter is likely to display values to within ±5%, whereas the digital multimeter will depend on the resolution provided by the A/D conversion process and number of digits on the display, but has a typical resolution of 0.1Ω and quoted basic accuracy of better than ±1%.

Appendix B - A Bit of Theory

Section 1: Noise

THIS CAN BE defined as any unwanted disturbance that is superimposed on a wanted signal. It will interfere with the information contained within the wanted signal and in the limit will prevent it from being decoded. In radio reception it can produce crackling or hissing, which in the limit will mask the desired speech or music. In data transmission it will affect the reliability with which data is decoded creating a much higher error rate.

Noise originates from a wide variety of sources, both man-made and natural. At frequencies up to around 21MHz, external noise is generally greater than noise generated within a receiver. Above this frequency and up into microwaves, receiver generated noise is generally dominant, especially at 144MHz and above.

Receiver noise is generated by both passive and active devices. There are many types of noise, and certain types, such as thermal and shot noise, cannot be avoided because they are a fundamental property, arising from the temperature of, and the current flow through, the component respectively. Moreover, in a collection of otherwise identical components (such as transistors), some may generate less noise than others, which is why these preferred devices may be classified as 'low-noise' in the literature. This variation originates from unavoidable manufacturing defects in the semiconductor at the atomic level, which produce extra energy levels. These unwanted levels in turn generate 1/f (and other) noise, which is frequency dependant and gives rise for example to the unwanted close-to-carrier noise on a local oscillator. Current passing through a semiconductor also generates shot noise proportional to the current flow and reflects the fact that the DC current is not actually constant, but is a stream of discrete charges (holes and electrons) passing through the device. Thermal noise in contrast, is generated within components (such as resistors) by the random movement of charge and results in a random voltage. At absolute zero this voltage is zero, increasing as temperature rises.

If a group of components in the front-end of a receiver generate noise in the µV region, this can well mask received signals of the same order. It is therefore important that receiver front-ends are well designed, use low-noise devices in key positions, and are set up in order to minimise the effect of noise.

For measurement purposes, 'white noise', with its broadband frequency-independent spectrum is particularly important for making measurements. This is generated by semiconductors as shot noise, and by resistors as thermal noise. The former is important as a noise source while the latter helps to determine the system 'noise floor' and hence the smallest discernible signal.

Thermal Noise (P_r)

For a resistor R at a Kelvin temperature of T, the noise power generated in a bandwidth of B Hertz is:

$P_r = kTB$ where k=Boltzmann's Constant (1.38×10^{-23} Joules per degree Kelvin)

Thus at 298 Kelvin (~25°C), this gives the well-known result:

$P_r = -174$dBm/Hz

Added Noise (P_a)

Suppose a 50Ω load is connected to the input of an amplifier. The noise power (P_{oAmp}) coming from the amplifier's output will be that of the resistor (P_r), multiplied by the gain of the amplifier (G), plus some noise power added by the amplifier (P_{amp}); hence:

$P_{oAmp} = GP_r + P_{amp}$

Excess Noise Ratio (ENR)

But if the load resistor is replaced by an aerial, and there is more noise at the input (by a factor N), the new power level at the output is:

$P_{oAerial} = NGP_r + P_{amp}$

The output has thus increased by the ratio:

$Y = P_{oAerial}/P_{oAmp}$

Where this noise 'on/off' ratio is known as the Y-factor.

Solving the above equations for the Excess Noise Ratio (ENR) gives:

$ENR = Y[1 + P_{amp}/GP_r] - P_{amp}/GP_r$

Clearly, if the noise ratio (Y) 'with and without the aerial' is high, and the amplifier added noise is insignificant, the terms in P_{amp} become negligible, and so the formula simplifies to ENR=Y.

Noise Factor

If this is not the case, and amplifier noise dominates (for example with a microwave amplifier), the amplifier's noise factor (F) must be taken into account. This may be derived from its Noise Figure (NF quoted in dB), since:

$F = 10^{0.1NF}$

Now the added noise may be expressed as:

$P_{amp} = (F-1) GP_r$

Which when substituted in the equation for ENR gives:

$ENR = YF - F + 1$

Or in measurable terms;

$ENR = (V_{outAerial}/V_{outAmp})^2 F - F + 1$

→ For more detail see 'Measuring Spectrum Pollution at VHF', John Worsnop, G4BAO, *RadCom*, January 2018, citing Alwyn Seeds, G8DOH, at the 2017 RSGB Convention.

Noise Figure and Noise Temperature

It is worth noting that F is the ratio of the Signal-to-Noise at the input compared to the output; that is:

$F = SNR_{in}/SNR_{out}$

And

$NF = 10log_{10}(F)$

Ideally F =1 (noiseless), but if in reality say F=2, then NF=3dB

Sometimes amplifiers are specified in terms of noise temperature (T), which is related to Noise **Figure by** the expression:

$NF_{dB} = 10Log_{10}(1 + T/290)$

Hence a 3dB noise figure, gives the result T=290K.

For cascaded stages, the overall Noise Factor is given by the series:

$F = F_1 + \{(F_2-1)/G_1\} + \{(F_3-1)/(G_1G_2)\}....$

However, stages beyond the second generally contribute a negligible amount to the noise figure, so terms beyond the second are neglected.

Minimum Discernible Signal (MDS)

This is set by the sum of the power in the noise floor and the detection bandwidth B (Hz), and is given by:

$MDS = -174dBm + NF + 10Log_{10}(B)$

→ For a discussion of the above in the context of building a low-noise amplifier for 23cm, see: 'Homebrew', Eamon Skelton, EI9GQ, *RadCom*, September 2016.

Section 2: Transmission Lines

Transmission lines and antenna systems have various characteristics, typical of which are:

- Antenna system impedance
- Antenna current
- Antenna resonance and bandwidth

Reflection coefficient, return loss, and VSWR are all measurements made on antenna systems, as well as on the input and output of amplifiers, filters, matching networks etc. They are related to each other and this section gives some idea of the relationships.

Antenna equivalent circuit

An antenna, or antenna with transmission line, presents an impedance to whatever is driving it. At a particular frequency, this impedance can be represented by a series or parallel circuit as shown in **Fig B1.1**.

$R_p = (R_s^2 + X_s^2)/R_s$ and $X_p = (R_s^2 + X_s^2)/X_s$

$R_s = R_p X_p^2 /(R_p^2 + X_p^2)$ and $X_s = X_p R_p^2/(R_p^2 + X_p^2)$

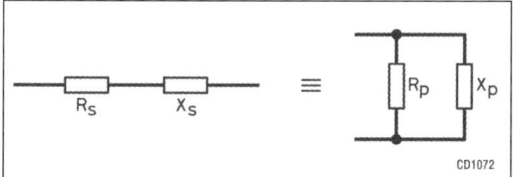

Fig B1.1 Series and parallel equivalence

For optimum power transfer, the resistive part should equal the source resistance and the reactive part should cancel the source reactance - in effect the condition for resonance.

Do not forget that what the transmitter actually 'sees' the impedance at the antenna modified by the intervening transmission line. If, and only if, this transmission line is a multiple of a half wavelength (taking into account the cable velocity factor) will the impedance at the transmitter be that of the antenna.

Reflection Coefficient, Return Loss and VSWR

If the power transmitted by the source is PF watts, and the power reflected back is PR watts, then the Return Loss (RL) in dB is given by:

$RL = 10\log_{10}(P_F/P_R)$

Or in other words,

$RL = P_F(dBW) - P_R(dBW)$

Where PdBW is the power level expressed in decibels relative to 1W

A return loss of 30dB means that only a thousandth of the power is being returned, whereas 0dB means that it's all coming back!

Ideally, the load impedance Z_L should be resistive (no reactive part) and the same as the characteristic impedance of the transmission line Z_0 so that no power is returned, the return loss is infinite and the VSWR 1:1. In reality however, there will be some impedance 'mismatch' which leads to a reflection. The relationship between this impedance mismatch and the strength of reflection is given by the reflection coefficient ρ (rho), where:

$|\rho| = \left|\dfrac{Z_L + Z_0}{Z_L - Z_0}\right|$

Where | | means 'the absolute value of' - that is a negative number is still treated as if it were positive. [Note: When using S-parameters, ρ is also equivalent to S_{11}].

Now when we go to measure (for example) the mismatch between our transmitter and its antenna, what our meter actually tells us is the VSWR. This is defined as:

$VSWR = \dfrac{1 + |\rho|}{1 - |\rho|}$

Where $|\rho| = 10^{-RL/20}$ (since RL $= -20\log_{10}|\rho|$)
The important characteristic of this expression is that the bigger the reflection, the bigger the VSWR, or conversely, as the reflection $|\rho|$ gets smaller, so the VSWR approaches 1.

In terms of impedances, if $Z_L >> Z_0$, then the VSWR $\approx Z_L/Z_0$. If $Z_L << Z_0$, then VSWR $\approx Z_0/Z_L$.

Where >> means 'much greater than'; << means 'much less than' and ≈ means 'approximately equal to'.

Putting all of this together, a return loss of 30dB means 1/1000th of the power is returned and equates to a VSWR of 1:1.07; 20dB equates to 1:1.22 and 10dB to 1:1.9. The concept of return loss is also useful when measuring filter responses and when we want to know how well a filter matches the characteristic impedance (Z_0) of the system.

→ For a short tutorial and a Power/VSWR/Return Loss converter/calculator, see:
http://www.giangrandi.ch/electronics/anttool/swr.shtml

→ For the relationship between power, voltage, and the decibel (dB) see:
http://www.giangrandi.ch/electronics/anttool/decibel.shtml

Transmission Lines

For an ideal lossless transmission line of length l, and characteristic impedance Z_0, terminated in an impedance ZL, the input impedance to the line is given by:

$Z_{in} = Z_0[(Z_L + jZ_0\tan\beta l)/(Z_0 + j Z_L\tan\beta l)]$
where $\beta = 2\pi/\lambda$

An important property of a transmission line

is to convert, change, or 'transform' impedance according to its length.

Open- and Short-circuit stubs

If $Z_L=\infty$ (for example in an idealised open-circuit matching stub) then the above equation reduces to:

$Z_{in} = -jZ_0\mathrm{Cot}\beta l$

Similarly, if $Z_L=0$ (for an idealised short-circuit matching stub)

$Z_{in} = jZ_0\mathrm{tan}\beta l$

This means that a length of transmission line, which could be fabricated in for example coax, open-wire, waveguide, or microstrip, can be used to present (usually a shunt) impedance for matching purposes.

Since Z_{in} contains no resistive part this means that it presents a pure reactance, and we know that the reactance of a capacitance is given by $X_c=1/j\omega C$ and that of an inductance by $X_L=j\omega L$ where $\omega=2\pi f$. Hence substituting in the above:

For an open-circuit stub: $-jZ_0\mathrm{Cot}\beta l = X_c=1/j\omega C$

For a short-circuit stub: $jZ_0\mathrm{tan}\beta l = X_L=j\omega L$

That is:

$C = (\mathrm{tan}\beta l)/\omega Z_0$

$L = Z_0(\mathrm{tan}\beta l)/\omega$

In the specific case of very short lengths of line, $\mathrm{tan}\beta l \approx \beta l$, and the above simplify down to:

$C \approx l/(\lambda Z_0 f)$

$L \approx Z_0 l/f\lambda$

Up to a quarter-wavelength long, a transmission line with a short at one end presents an inductive reactance at the other. In contrast, an open will present a capacitive reactance.

This is the basis of matching – creating the impedance required to make the circuit work properly. Often this is about maximum power transfer or minimising noise figure.

Quarter-Wave Lines

If length of transmission line is specifically a quarter wavelength long ($\lambda/4$), then it has special properties:

If $Z_L=0$, then $Z_{in}=\infty$ (a short circuit load presents infinite input impedance);

If $Z_L=\infty$, then $Z_L=0$ (an open circuit load presents zero input impedance)

This property can be very handy when designing bias circuits at microwave and millimetric frequencies. For example; to head-off the possibility of oscillation in a microstrip amplifier, RF can be prevented from entering or leaving a bias supply by using a disc roughly a third of a circle in circumference and $\lambda/4$ in radius. Ignoring 'fringing effects' this creates (ideally) $Z_L=0$ at the centre. If the bias-line from the centre is $\lambda/4$ (or $\lambda/4 + n\,\lambda/2$) long, then the impedance presented to the main transmission line is $Z_L=\infty$. This minimises loading of the bias circuit on the transmission line, which usually has a Z_0 in the order of tens of ohms, and the mismatch deters RF from entering the bias circuit.

Quarter-Wave Transformers

If the input impedance is purely resistive, then the output impedance will also be purely resistive, and vice versa. The relationship is given by:

$Z_0^2 = Z_{in}\,Z_L$

This is the basis of the 'quarter-wave transformer' found particularly in higher frequency circuits. It provides a *series* impedance, whereas a stub provides a *parallel* impedance.

Half-Wave Lines

The significance of a half wave line is that the impedance at one end is the same as the other. This can be useful where some extra physical distance is required to generate a specific impedance as n $\lambda/2$ lengths can be inserted. However line-loss and dispersion is increased and this may in turn have an adverse effect on noise figure and bandwidth.

Smith Chart

A 'plain' Smith chart (**Fig B1.2**) will show lines of constant resistance and constant reactance, with 'distance' (electrical phase) in terms of wavelength or angle from the source or load being displayed around the perimeter. This allows manual plotting of complex imped-ance as a function of frequency, as well as analysis of the effects of perhaps changing the value of a lumped element, the length of a transmission line, or adding a shunt stub. Mismatches and VSWRs can be easily seen and information deduced.

In skilled hands the Smith chart is an in-credibly useful aid to high frequency design, matching, and circuit analysis. When the chart is observed as a polar plot on a net-work analyser it reveals a lot about what the circuit is doing as a function of frequency -

for example a parasitic resonance shows up as a circle on the main locus at its resonant frequency. Where a mismatch is being cre-ated deliberately, for example in a low-noise amplifier to generate the correct impedance for best noise figure from a device, the ef-fect of the mismatch on circuit behaviour with frequency (such as gain response) can be studied.

Professional-grade design and analysis pro-grams with a Smith chart interface contain device libraries and equivalent circuits for all manner of elements (including bond-wires

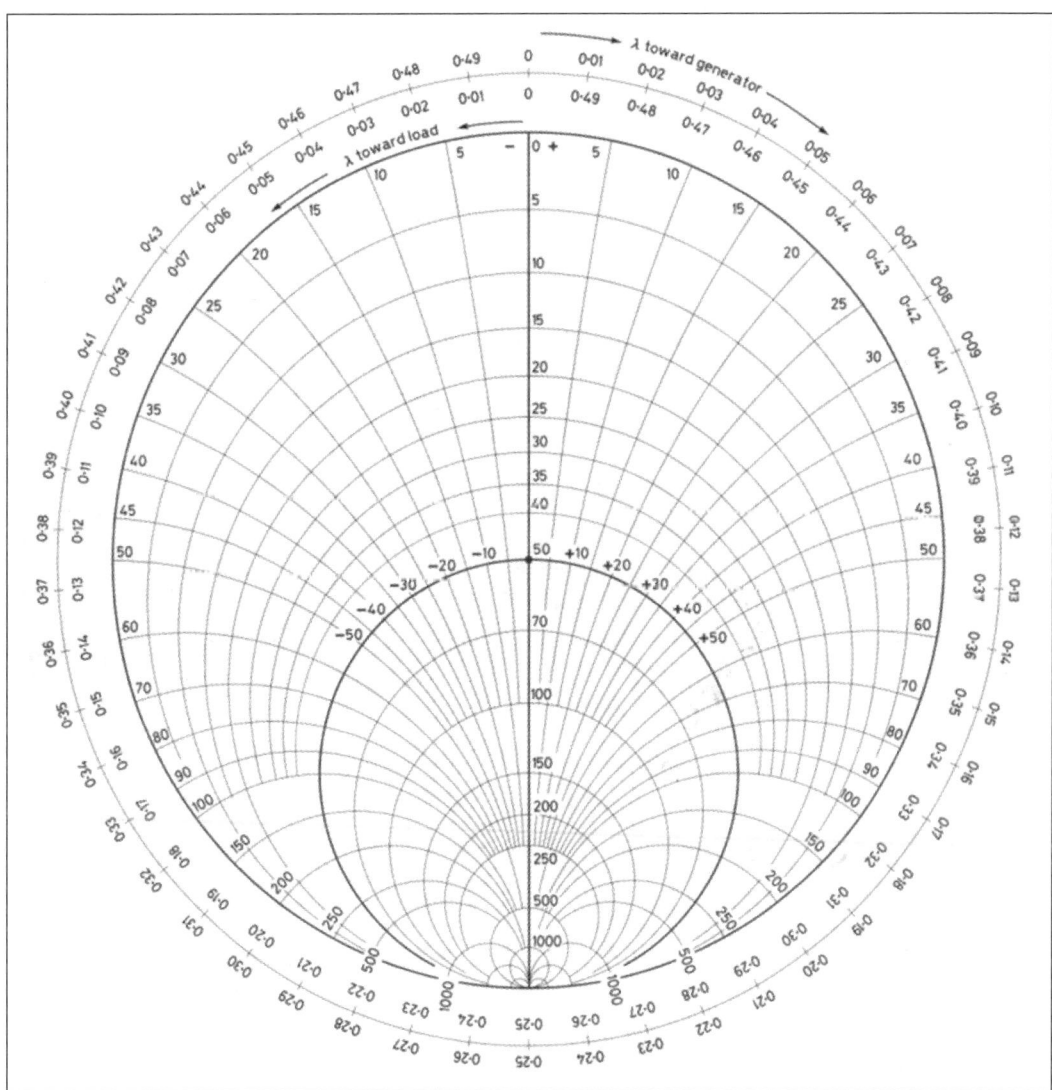

Fig B1.2 The Smith chart – the engineer's complicated but very helpful friend

and discontinuities). These are beyond the scope of most Radio Amateurs; however, there are Smith chart programs that will plot the response of basic circuits, such as a filter, given its elements. A couple of examples are:

→ 'JJSmith' a Windows 10 compatible program.

http://www.tonnesoftware.com/jjsmith.html

→ 'SimSmith' by AE6TY for Windows 10, OS, and Linux. *http://www.ae6ty.com/smith_charts.html*

→ For real experts who are not prone to confusion, charts which show lines of conductance and susceptance (ie the complex admittance 'Y') at the same time as the complex impedance 'Z' may be had, with the lines in different colours (usually red and green). The latter can be handy when working with parallel elements (such as multiple matching stubs), where the admittances can be simply added together in the same way that series impedances can be added together. That said, it is very easy to convert impedance to admittance on a Smith chart and so having both is not usually necessary and is probably best avoided!

Filters

There are numerous programs that can help with the design and theoretical test of filters and attenuators to ensure that the expected in-band and out-of-band transmission and impedance responses are obtained. The following are examples of Windows 10 compliant programs for electrical filter design and analysis:

→ 'RFSim99'; *http://www.ad5gg.com/2017/04/06/free-rf-simulation-software/*

→ 'Elsie'; *http://www.tonnesoftware.com/elsie.html*

Section 3: Power & Relationships

Units of Power Measurement

Although the watt is the standard unit of measurement (and its divisions mega, kilo, milli, micro etc), RF engineers often find dBm and dBW much more convenient for expressing power levels and doing mental calculations.

• dBm - is a power expressed in decibels relative to one milliwatt.

• dBW - is a power expressed in decibels relative to one watt.

The relationship between these two is: dBW = dBm-30

Table B1.1 shows a range of equivalents between watts, dBW and dBm. From this it can be seen that a level of 10dBm is 10dB above one milliwatt, ie 10mW. Similarly a power level of 20dBW is 100 times that of one watt, ie 100 watts.

→ To further understand units for RF and microwave power measurements; see:

http://www.radio-electronics.com/info/t_and_m/rf-microwave-power-meter/basics-tutorial-introduction.php

Relationship between volts and dBm

For a 50Ω system:

dBm = dBμV -107, or in other words, dBm = $-107 + 20\log_{10}(\mu V)$

For a 75Ω system:

dBm = dBμV -108.7, or in other words, dBm = $-108.7 + 20\log_{10}(\mu V)$

→ Alternatively, use one of the online conversion sites or download a program.

Watts	dBW	dBm
1μW	-60	-30
10μW	-50	-20
100μW	-40	-10
1mW	-30	0
10mW	-20	10
100mW	-10	20
1W	0	30
10W	10	40
100W	20	50
1000W	30	60

Table B1.1 Equivalents between watts, dBW and dBm

Appendix C - Reference Data

SOME OF the more common forms of component identification and connection media are given below. Additional data can be found in many of the standard electronic and radio reference books and of course, on the internet

C1 Passive Component Identification Codes

The colour codes in **Table C1.1** are applicable to resistors, capacitors, and inductors and are occasionally used for small semiconductors such as diodes.

Colour	Number	Multiplier	Tolerance
Black	0	1	-
Brown	1	10	±1%
Red	2	100	±2%
Orange	3	1k	-
Yellow	4	10k	-
Green	5	100k	±0.5%
Blue	6	1M	±0.25%
Mauve	7	10M	±0.1%
Grey	8	-	-
White	9	-	-
Gold	-	1/10	±5%
Silver	-	1/100	±10%

Table C1.1 Colour codes for components

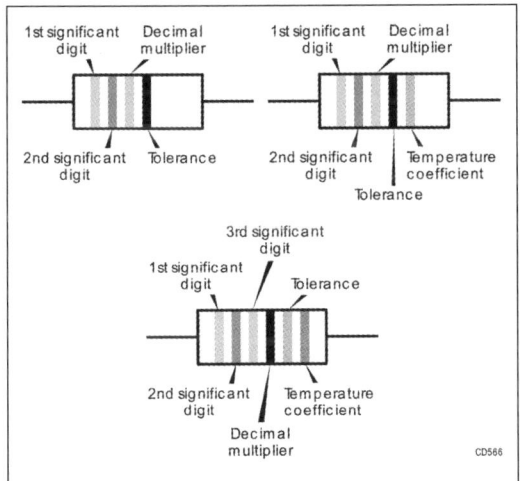

Fig C1.1 Resistor colour banding

Resistors

These use either colour coding or an alpha-numeric code

Colour coding comes with four or five colour bands (**Fig C1.1**). If there is no tolerance band, ±20% is assumed. A sixth band indicates temperature coefficient as follows:-

Brown	100ppm	Orange	50ppm
Red	50ppm	Yellow	25ppm.

Alphanumeric - this should follow BS1852 (IEC publication 62) as in these examples:

0.22Ω is R22	220Ω is 220R
1.0Ω is 1R0	2.2kΩ is 2K2
2.2Ω is 2R2	22kΩ is 22K
22Ω is 22R	1MΩ is 1M
100Ω is 100R	2.2MΩ is 2M2

A letter denotes the resistor's tolerance:-

F = ±1%	K = ±10%
G = ±2%	M = ±20%
J = ±5%	Z = -20% +80%

Hence a resistor marked 4K7J would be 4.7kΩ, ±5%. The colour coding for this would be yellow, mauve, red, gold.

Resistors are made in various ranges of values classified to BS2488 (IEC publication 63). The most common are defined in **Table C1.2**.

E6	10	15	22	33	47	68						
E12	10	12	15	18	22	27	33	39	47	56	68	82
E24	10	11	12	13	15	16	18	20	22	24	27	30 33
	36	39	43	47	51	56	62	68	75	82	91	

Table C1.2 Resistor ranges

Capacitors

These are generally alphanumerically coded, although attempts have been at colour coding. All schemes are included here for reference.

• pF, nF and µF, may be shortened to p, n and µ.

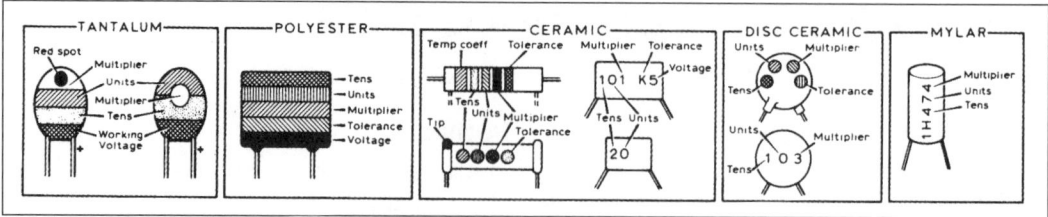

Fig C1.2 Various forms of capacitor coding

• In a three digit code based on the pF, the first two digits are the significant numbers and the third the 'multiplier' (power of 10). Hence 102 = 10x10^2= 1000pF and 473 = 47x10^3 = 47nF.

There is a colour coding on the top of some ceramic plate capacitors defined as follows:-

Green	High K
Yellow	Medium K
Red/Mauve	Low K+100ppm/°C
Black	Low K±0ppm/°C NPO
Orange	Low K-150ppm/°C N150
Mauve	Low K-750ppm/°C N750
Orange/	
Orange	Low K-1500ppm/°CN1500

Apart from the tolerance coding under resistors two more also apply for capacitance:-

C = ±0.25pF   D = ±0.5pF

If capacitors are colour coded then they normally refer to either a polyester type or to tantalum. For polyester, read the value in pF, for tantalum bead read in μF. **Fig C1.2** shows how the colour bands are organised. The colour coding for voltage rating is shown in **Table C1.3**.

Colour	Tantalum	Polyester
Black	10V	-
Brown	-	100V
Red	-	250V
Yellow	6.3V	400V
Green	16V	-
Blue	20V	-
Grey	25V	-
White	3V	-
Pink	35V	-

Table C1.3 Colour coding of the voltage rating of tantalum and polyester capacitors

Inductors

Inductors may carry alphanumeric or colour coding.

Alphanumeric coding is based on μH and arranged as a three digit code as for resistors and capacitors. The only additional code that may occur is an 'n' for nH (less than 100nH). From 100nH up to 1μH the code will be written as 'R22' for 0.22μH. From 1μH up to 10μH the code will be as '8R2', for 8.2μH. After this it becomes the number and multipliers. Also used may be the unit placed where the decimal point should be For example μH22 is 0.22μH, 5mH6 is 5.6mH.

If a colour coding is used (**Fig C1.3**), the first band is a broad silver identifier for inductor. The next two bands are the significant numbers followed by the multiplier band and then a tolerance band. The base unit is the μH. In some inductors a gold band may be used at the position of the decimal point.

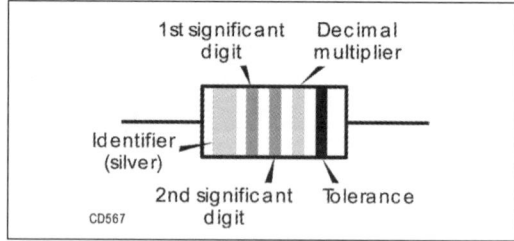

Fig C1.3 Inductor colour banding

C2 Semiconductor Identification Codes

There are various systems of coding discrete semiconductor devices; some are manufacturer dependent, others attempt to follow one of the three forms of coding explained below.

The Pro-Electron numbering System

This is the system most widely used in Europe. The coding consists of two letters followed by a serial number. The serial number consists of either three digits, or of one letter plus two digits, e.g. BC109 or BLX98.

The first letter is defined as follows:-

A germanium
B silicon
C compound materials such as gallium arsenide
D compound materials such as indium antimonide
R compound materials such as cadmium sulphide

The second letter denotes the general function of the device as follows:-

A detection diode, high speed diode, mixer diode
B variable capacitance diode
C transistor for AF applications (but not power type)
D power transistor for power applications
E tunnel diode
F transistor for RF applications (but not power type)
G multiple of dissimilar devices; miscellaneous devices
L power transistor for RF applications
N photo-coupler
P radiation sensitive device such as a photo-diode, photo-transistor, photo-conductive cell or a radiation detector diode
Q radiation generating device such as an LED
R controlling/switching device (e.g. thyristor) having a specified breakdown characteristic (but not a power type)
S transistor for switching applications (not power types)
T controlling and switching power device (e.g. thyristor) having a specified breakdown characteristic
U power transistor for switching applications
X multiplier diode such as varactor or step recovery diode
Y rectifier diode, booster or efficiency diode
Z voltage reference or voltage regulator diode, transient suppressor diode

The remainder of the type number is a serial number indicating a particular design or development and is one of the following two groups:

1. Device intended primarily for use in consumer applications (radio and TV, audio amplifiers, tape recorders, domestic appliances etc). The serial number consists of three figures.

2. Device intended mainly for non-consumer applications, e.g. industrial, professional and transmitting equipment. The serial number consists of one letter (Z,Y,X,W etc) followed by two figures.

Where there is a range of variants of a basic type of rectifier diode, thyristor or voltage regulator diode, the type number as defined above is often used to identify the range; further letters and figures are added after a hyphen to identify individual types within the range. These additions are as follows:-

Rectifier diode and thyristors: The group of figures indicates the rated repetitive peak reverse voltage VRRM or the rated repetitive peak off state voltage VDRM, whichever value is lower in volts for each type. If there is a final letter R, this denotes a reverse polarity version, I.e. stud is anode.

Voltage regulator diodes, transient suppression diodes: The first letter indicates the nominal percentage tolerance in the operating voltage VZ as follows:-

A = ±1%	B = ±2%	C = ±5%
D = ±10%	E = ±15%	

The letter is omitted on transient suppressor diodes. The succeeding group of figures indicates the typical operating voltage V_Z for each type at the nominal operating current I_Z rating of the range. For transient suppressor diodes the figure indicates the maximum recommended stand-off voltage V_R. The letter V is used to denote the position of the decimal point.

In addition to the above, small signal devices sometimes have a letter after them to define an h_{fe} range as follows:

A - h_{fe} between 125 and 260

B - h_{fe} between 240 and 500

C - h_{fe} between 450 and 900

The JEDEC System

A system deriving from the USA, devices can be registered and allocated a coding such as 1Nxxxx, 2Nxxxx etc, as follows:-

1st symbol	a number 1,2,3 etc
2nd symbol	a single letter N
3rd symbol	the device's registration number (xxxx)

Where the 1st symbol is defined as follows:-

1   a two-pin device such as a diode or rectifier

2   a three-pin device such as a transistor or thyristor

3   a four-pin device such as a dual gate MOSFET

The Japanese Industrial Standard (JIS-C-7012)

This designation code is made from five symbols:-

The 1st symbol indicates the type of semiconductor:

0   phototransistor

1   signal, rectifier, or varactor diode

2   bipolar, junction FET (single gate), thyristor etc.

3   as 2 but with two gates

The 2nd symbol (a letter S) indicates that the device is diffused from semiconductor elements.

The 3rd symbol indicates polarity and application of the device:

A   high frequency PNP transistor

Numbering	Family	Comments
74xx	Standard TTL	Standard
74ACxx	Advanced CMOS	high speed, reduced power, CMOS voltage levels
74ACT	Advanced CMOS	as AC type but TTL input voltage levels
74AHC	Advanced CMOS	high speed, CMOS logic levels
74AHCT	Advanced CMOS	high speed, TTL logic levels
74ASxx	Advanced Schottky TTL	improved speed LS type
74ALSxx	Advanced LS TTL	twice the speed, half the power of LS TTL
74Cxx	CMOS	low power, high noise immunity
74Fxx	Fast TTL	High speed switching
74HCxx	High speed CMOS	low power, speeds similar to LS TTL, CMOS voltage levels
74HCTxx	High speed CMOS	as HC type but TTL levels
74LSxx	Low power Schottky TTL	same speed as standard but one fifth power.
74Sxx	Schottky TTL	faster than LS TTL, takes more power
74VHC	CMOS	very high speed. CMOS input
74VHCT	CMOS	as 74VHC with TTL input and output
4xxx/45xx	CMOS	low power, high noise immunity, 3-15V operation, medium speed
1000xxx	ECL	for use in very high speed systems

Table C2.1 Logic IC families

B   low frequency PNP transistor
C   high frequency NPN transistor
D   low frequency NPN transistor
F   P-gate thyristor
G   N-gate thyristor
J   P-channel FET
K   N-channel FET

The 4th symbol (made of two or three digits) has a value of 11 or more given according to the sequence of registration with JIS.

The 5th symbol (a letter) indicates a revision of the product originally introduced.

→ If the 1st symbol is 0 or 1, only numbers are used for the 3rd symbol.

Logic families

A multitude of logic ICs are, or have been available. **Table C2.1** attempts to denote some family types:

The 74 series are usually pin-for-pin compatible with other members of the family, but the characteristics of the IC must be considered if trying to substitute; i.e. operating levels (CMOS or TTL), operating speeds, drive currents etc. The 74HC4xxx series are pin-for-pin compatible with the 4000 series CMOS ICs. The 1000 series ECL devices are no longer in production.

C3 Surface Mount Device (SMD) Identification Codes

There is an abundance of device codes and size codes which makes correctly interpreting component markings (if any!) a nightmare. The following is provided as a guide:

Resistors

There are various case sizes for SMD resis-

tors, each with a 4-digit number; the first two digits represent the length, and the second two the width in inches. **Table C3.1** shows some of the case sizes.

Table C3.1 Typical SMD resistor case sizes

The power rating quoted is typical as it can vary between manufacturers. There are variations on a theme depending on the technology of the device. In addition, other sizes and shapes have involved, mainly due to power dissipation.

Values: Many SMD resistors do not have any markings on them to indicate their value, so will need to be measured. If they do have markings, three figures are used - the first two indicate the significant figures, and the third is a multiplier. Thus the figures '473' imply a resistance of 47×10^3 ohms or 47kΩ.

→ Beware resistors marked with figures such as 100. This is not 100 ohms, but it follows the scheme exactly represents 10×10^0, ie 10x1 = 10 ohms. Sometimes these resistors are marked with just two figures, i.e. 10 to prevent any misunderstandings.

Tolerance: Because SMD resistors are manufactured using mainly metal oxide film, they are available in relatively close-tolerance values. Normally 5%, 2%, and 1% are available. For specialist applications, 0.5% and 0.1% values can be obtained.

Temperature coefficient: Again, because of the use of metal oxide film, SMD resistors have a good temperature coefficient. Values of 25, 50 and 100 ppm/C are usually available.

Capacitors

Capacitors up to 1µF are available in the same case sizes as resistors, with the most

Size Code	Typical pwr rating	Dia (in)	Dia (mm)
0402	0.0625W	0.04 x 0.02	1.0 x 0.5
0603	0.0625W	0.06 x 0.03	1.6 x 0.8
0805	0.1W	0.08 x 0.05	2.0 x 1.25
1206	0.25W	0.12 x 0.06	3.2 x 1.6
2010	0.5W	0.20 x 0.10	5.0 x 2.5
2512	1W	0.25 x 0.12	6.4 x 3.1

Table C3.1 Typical SMD resistor case sizes

Case	Size (LxWxH) in mm
A	3.2x1.6x1.6
B	3.5x2.8x1.9
C	6.0x3.2x2.6
D	7.3x4.3x2.9
E	7.3x4.3x4.1

Table C3.2 Some typical tantalum capacitor case sizes

common type being ceramic. Above 1μF tantalum capacitors are the norm, but larger values may be aluminium electrolytic. Case sizes for tantalums are indicated by letters, and **Table C3.2** gives the code for some of the smaller sizes. A bar denotes the positive side of the electrolytic.

As for the value of the capacitor, it may be printed on, it may have a manufacturers code, or it may have nothing at all! Often SMD electrolytic capacitors are marked with the value and working voltage. There are two methods used. One is to include their value in microfarads (μF), and another is to use a code. In the first method, a marking of 33μ6V would indicate a 33μF capacitor with a working voltage of 6 volts. An alternative code system (**Table C3.3**) employs a letter followed by three figures. The letter indicates the working voltage as defined in the table below and the three figures indicate the capacitance in picofarads. As with many other marking systems the first two figures give the significant figures and the third, the multiplier. In this case a marking of G106 would

Letter	Voltage
e	2.5
G	4
J	6.3
A	10
C	16
D	20
E	25
V	35
H	50

Table C3.3 SMD electrolytic capacitor working voltage coding

indicate a working voltage of 4 volts and a capacitance of 10×10^6 picofarads or 10μF.

Inductors

Sizes tend to follow those for resistors but as always there are "special" sizes. They may have markings on, but physically may be difficult to distinguish from resistor. If in doubt, use a DMM or capacitance meter to check.

Semiconductors

There are a wide variety of surface-mount packages used for semiconductors including diodes, transistors and integrated circuits, and results from the large variation in the level of interconnectivity required. Some of the main packages are given below:

Transistor packages

• SOT-23 - Small Outline Transistor. This has three terminals for a diode or transistor but can have more pins when used for an integrated circuit such as an operational amplifier. It measures 3 x 1.75 x 1.3mm

• SOT-223 - Small Outline Transistor. This package is used for higher power devices. It measures 6.7 x 3.7 x 1.8mm. There are generally four terminals, one of which is a large heat-transfer pad

Integrated circuit packages

• SOIC - Small Outline Integrated Circuit. This has a dual in line configuration and gull wing leads with a pin spacing of 1.27 mm. The two most popular sizes are SOIC-8 and SOIC-14 (also named SO-8 and SO-14).

• TSOP - Thin Small Outline Package. Thinner than the SOIC, it has a smaller pin spacing of 0.5 mm

• SSOP - Shrink Small Outline Package. This has a pin spacing of 0.635 mm

• TSSOP - Thin Shrink Small Outline Package.

• PLCC - Plastic Leaded Chip Carrier. This type of package is square and uses J-lead pins with a spacing of 1.27 mm

• QSOP - Quarter-size Small Outline Package. It has a pin spacing of 0.635 mm

• VSOP - Very Small Outline Package. Smaller than the QSOP, and has pin spacing of 0.4, 0.5, or 0.65 mm.

• LQFP - Low profile Quad Flat Pack. This package has pins on all four sides. Pin spacing varies according to the IC, but the height is 1.4 mm

• PQFP - Plastic Quad Flat Pack. A square

plastic package with an equal number of gull-wing style pins on each side. Typically narrow spacing and often 44 or more pins. Normally used for VLSI circuits

• CQFP - Ceramic Quad Flat Pack. A ceramic version of the PQFP.

• TQFP - Thin Quad Flat Pack. A thin version of the PQFP.

• BGA - Ball Grid Array. A package with pads underneath to make contact with a printed circuit board. Before soldering, the pads appear as solder balls, giving rise to the name. By placing the pads underneath the package there is more room for them, thereby overcoming some of the problems of the very thin leads required for the quad flat packs. The ball spacing on BGAs is typically 1.27 mm

Device codes

→ An area fraught with difficulty and confusion! Try looking in the SMD Codebook by GM4PMK at:

www.marsport.org.uk/smd/mainframe.htm

C4 Coaxial Connectors Cable and Wire

Connectors: A list of some common coaxial connectors and their characteristics is given in **Table C4.1**. Adapters exist for conversion between the various types of connector; alternatively one can make a short conversion lead.

→ Note: Fitting a connector onto a cable *properly*, can be a challenge. For exploded views of some connectors and how they should be assembled, search on the Internet or consult the *VHF/UHF Handbook*, by Andy Barter, G8ATD, RSGB ISBN 9781-9050-8631-8, p297ff.

→ Care! - The common UHF connector (PL259/SO239) is available from many sources but its quality can be somewhat variable – examine the item, especially the insulation material. Try to purchase connectors from reputable suppliers.

European	American
UR43	RG58B/U
UR57	RG11A/U (stranded)
UR67	RG213
UR70	RG59B/U (solid core)
UR76	RG58C/U
UR95	RG174A/U (stranded)

Table C4.2 Coaxial cable equivalents

Cable: **Table C4.2** gives the approximate equivalents between the European UR and American RG series of coaxial cable.

Wire: Enamel copper wire (ECW) may be quoted in metric or wire gauge (SWG). **Table C4.3** gives the nearest equivalents.

For resistance per unit length and typical current carrying capacity it is easiest to look up tables on the Internet.

SWG metric (mm)	Dia. (mm)	Nearest
4	5.89	6.00
6	4.88	5.00
8	4.06	4.00
10	3.25	3.00
12	2.64	2.50
14	2.03	2.00
16	1.63	1.50
18	1.22	1.20
20	0.914	1.00
22	0.711	0.80
24	0.559	0.60
26	0.457	0.50
28	0.376	0.40
30	0.315	0.30
34	0.234	0.25
36	0.193	0.2

Table C4.3 SWG/metric wire equivalents

Connector Type	Impedance (Ω)	Typical upper frequency GHz	Typical peak voltage rating (V)	Notes
BNC	50 & 75	4GHz	500V	Bayonet coupling. Common on connecting leads for instruments
TNC	50 & 75	4GHz	500V	Screw coupling version of BNC type
N	50 & 75	10GHz	1kV	Screw coupling. UHF and VHF uses.
PL259 etc	not constant	200MHz	500	Screw coupling; typical for HF and video and some instrumentation
Min UHF	not constant	2GHz	450V	Screw coupling
PET100	50	1.5GHz	3kV	Screw coupling
F	75 (not constant)	1GHz	350V	Screw coupling for Satellite, TV cable etc
SMA	50	18GHz	450V	Screw coupling; high performance with extended frequency range.
SMB	50	4GHz	500V	Snap on;Suitable for use within equipment
TV	60-75	1GHz	Low power	Push fit; domestic TV connector
Min TV	50	1GHz	Low power	Miniature version of above connector

Table C4.1 Characteristics of various coaxial connectors

Appendix D – PCBs & Component Layouts

THIS APPENDIX CONTAINS PCB and component layouts for some of the projects in this book. The component layout is from the component side whilst the copper pattern is that seen from the copper side (unless stated otherwise).

In making the boards/masters, always take time to examine the board thoroughly before etching to ensure that there is no break in the copper tracks etc. If there is, the fault can be touched up with an etch-resist pen. Examine the board during etching to see if any bridges are being left: if there are, scratch through any etch-resist with a sharp pointed implement - e.g. a craft knife.

Disclaimer

While every endeavour has been made to ensure that these patterns are correct, neither the author, the originator, nor the RSGB can accept any liability for any mistakes or consequential damage. It is up to the constructor to ensure that any component used is fit for purpose.

Copyright Notice

It should be noted that the PCB patterns in this appendix are copyright. Permission is given only for the individual amateur to make a one-off for personal use. Commercial copying is expressly forbidden without licensing rights.

Chapter 8.2: Square-wave Generator 10Hz-1MHz

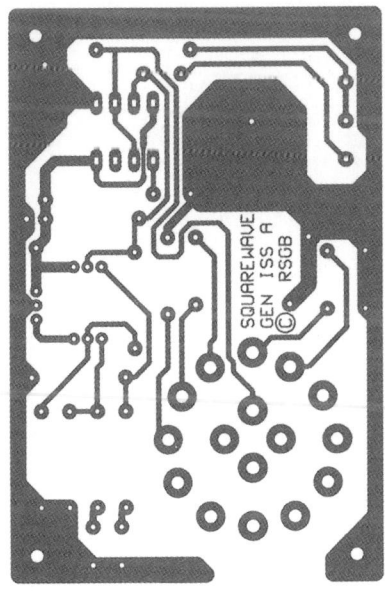

Pattern Fig D1.1

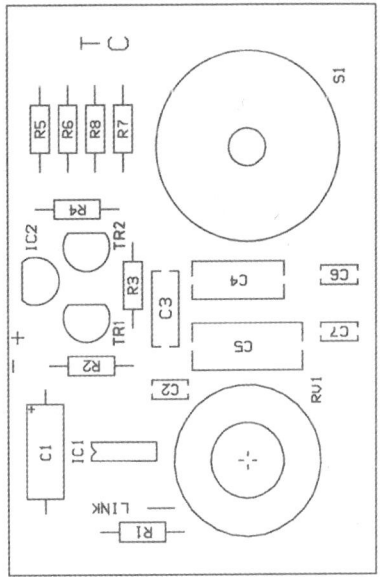

Layout Fig D1.2

Chapter 8.2: Sinewave Oscillator 10Hz-100KHz

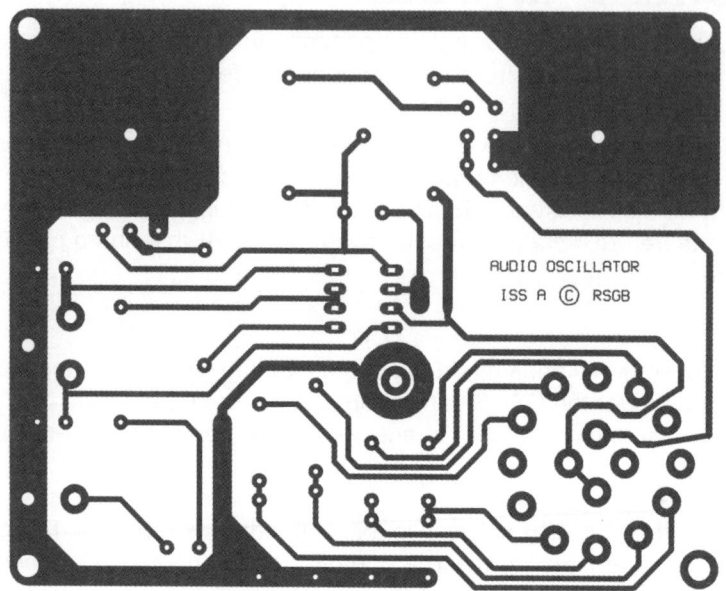

Pattern Fig D2.1

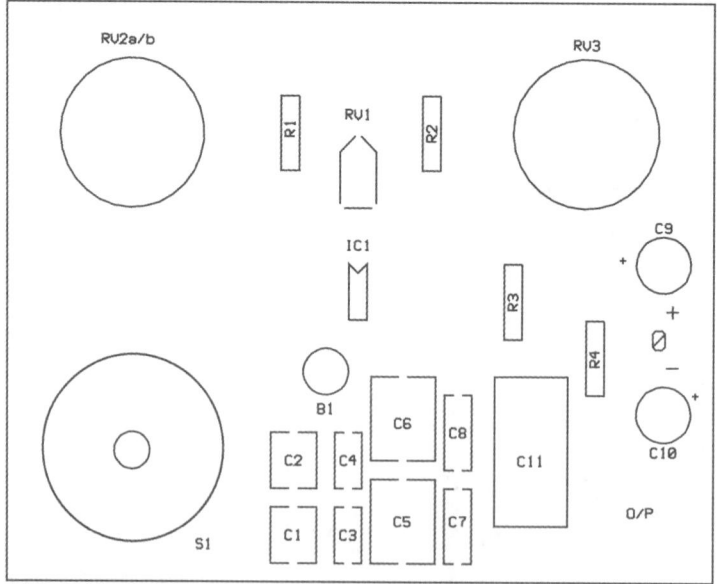

Layout Fig D2.2

Chapter 8.2: Two-Tone Oscillator

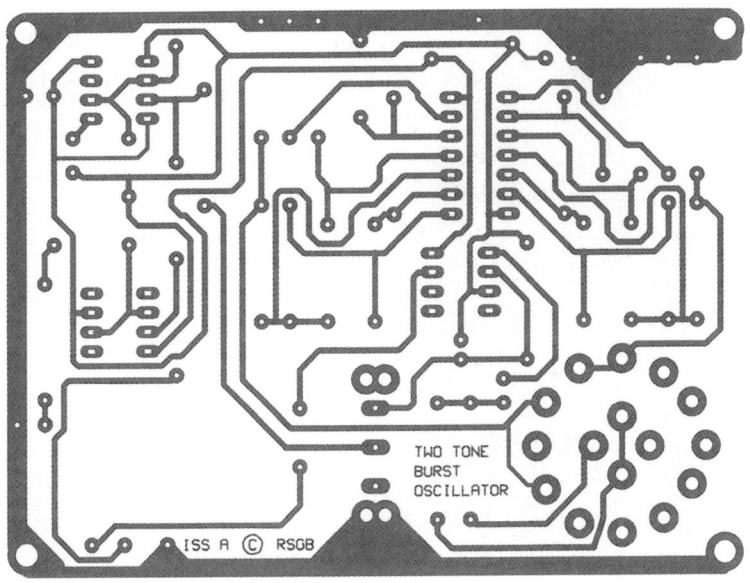

Pattern Fig D3.1

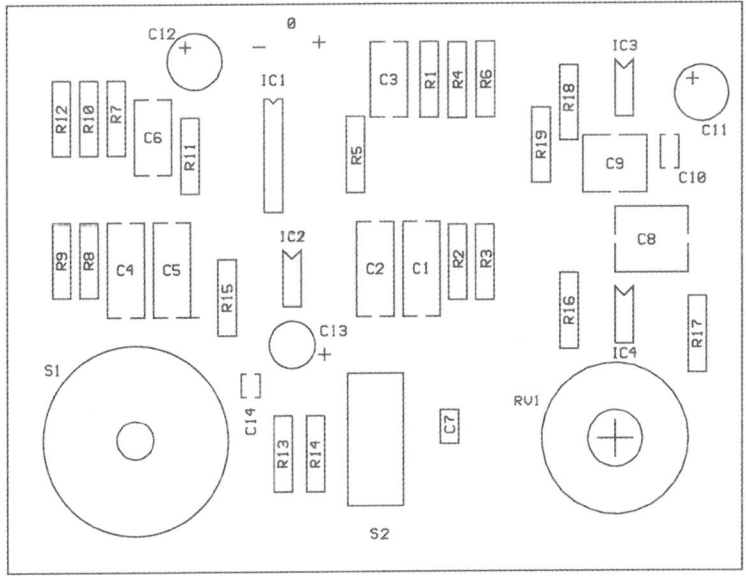

Layout Fig D3.2

Chapter 8.10: Noise Bridge for Measuring R and X

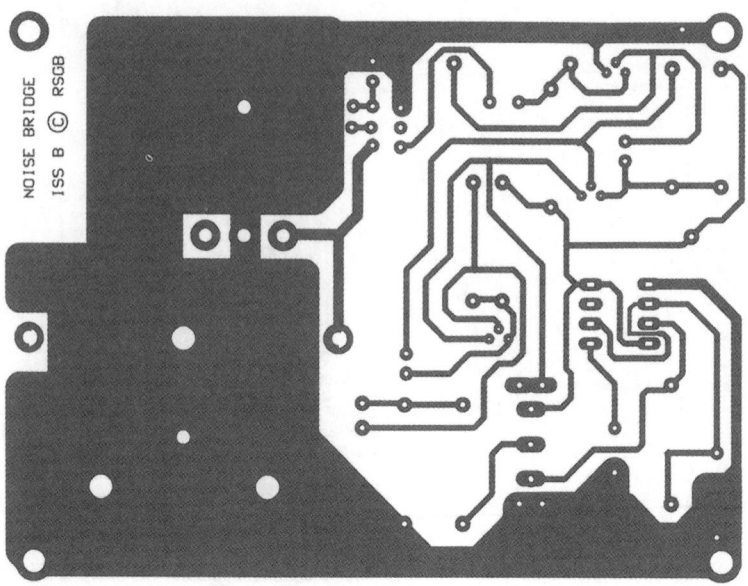

Pattern Fig D4.1

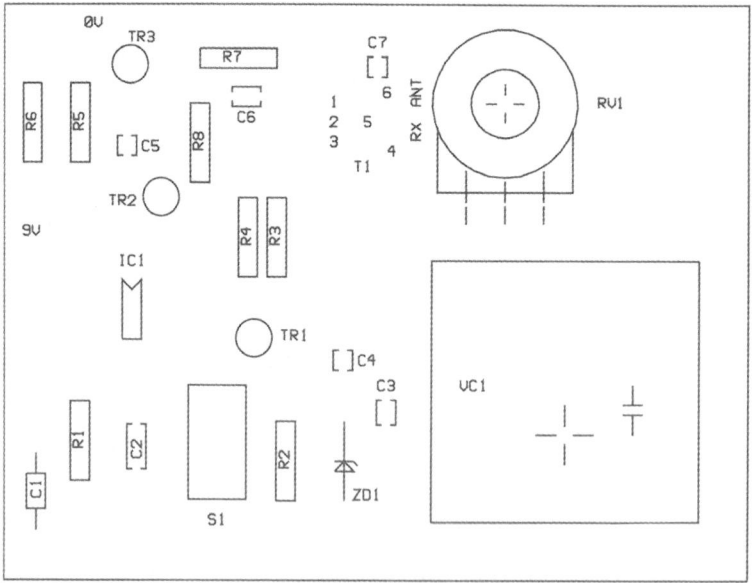

Layout Fig D4.2

Chapter 8.14: Triangle Wave Generator

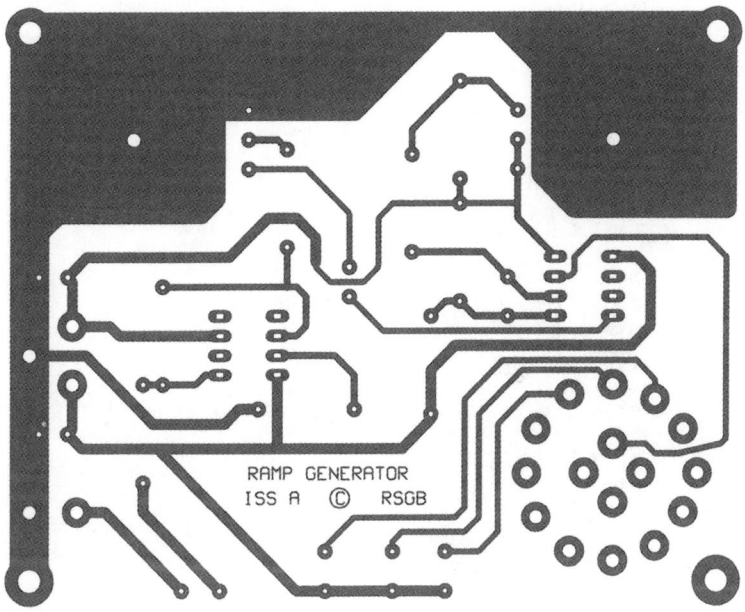

Pattern Fig D5.1

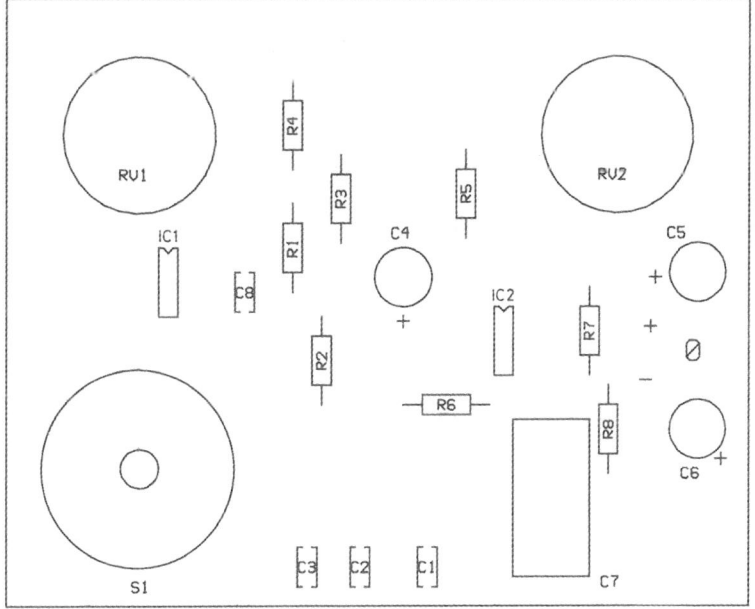

Layout Fig D5.2

Chapter 8.14: 440-550KHz Sweep Generator

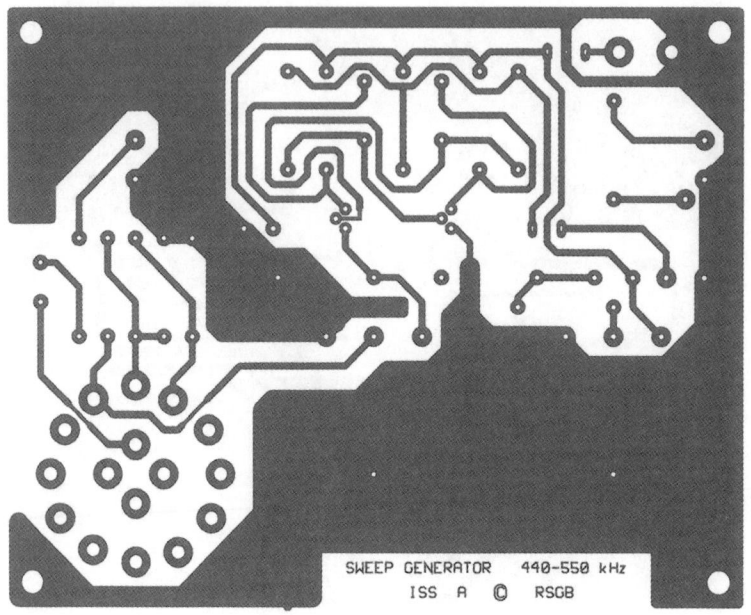

SWEEP GENERATOR 440-550 kHz
ISS A © RSGB

Pattern Fig D6.1

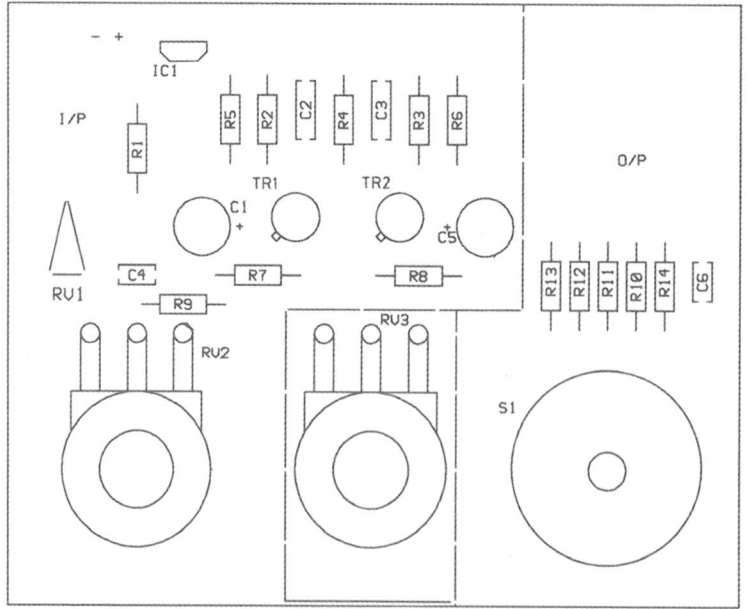

Layout Fig D6.2

Chapter 8.14: 9-10.7MHz Sweep Generator

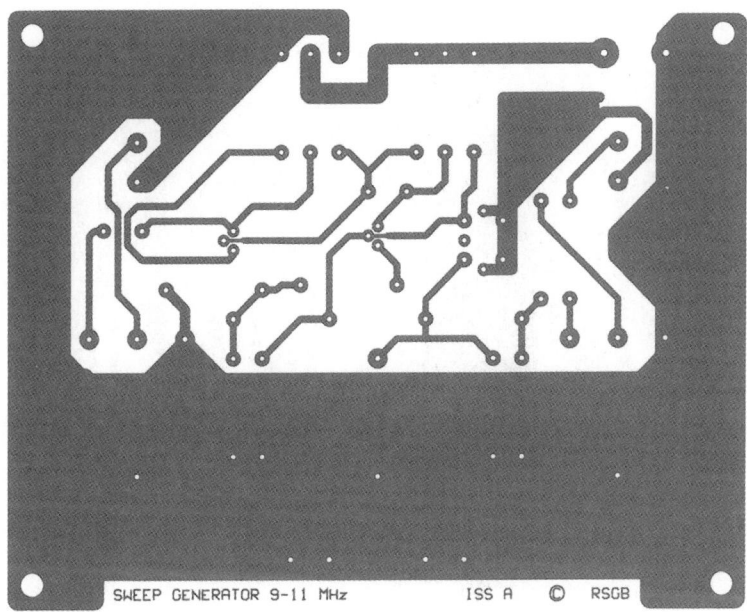

Pattern Fig D7.1

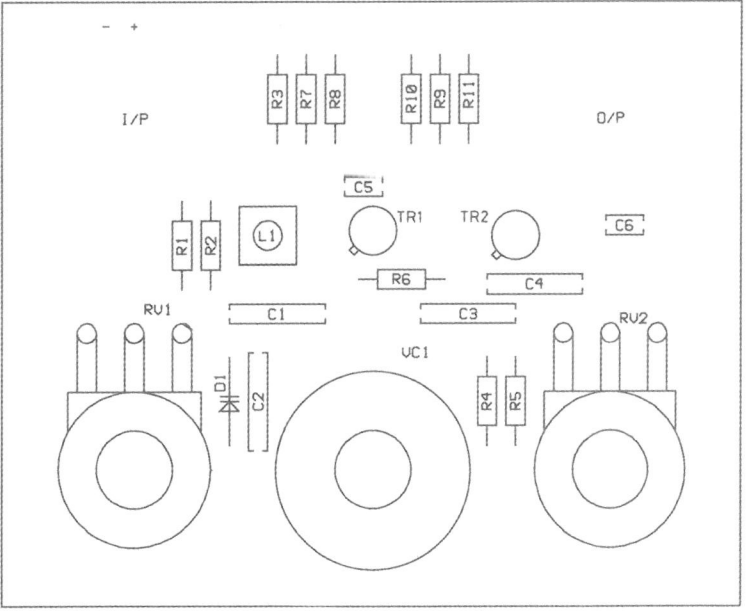

Layout Fig D7.2

Chapter 8.15: Fixed 9V, Low Current Power Supply

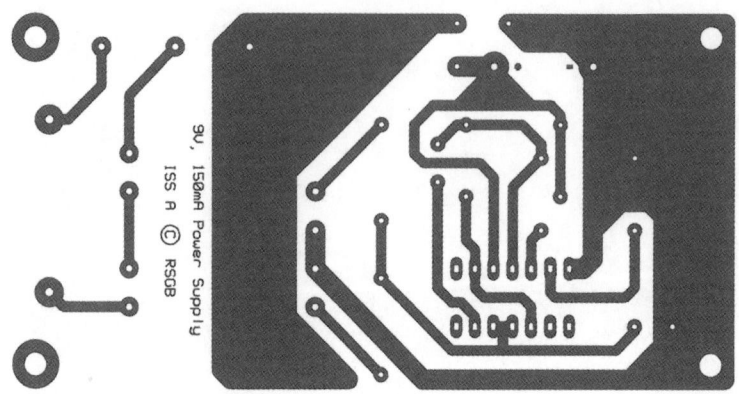

Pattern Fig D8.1

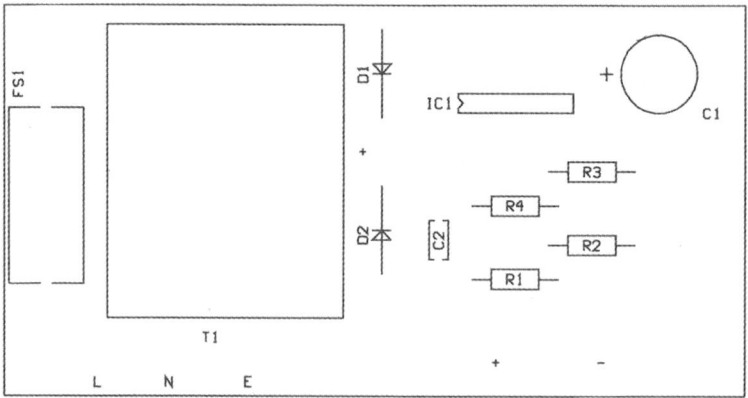

Layout Fig D8.2

Chapter 8.15: Limited Voltage Range 12V, 10A Power Supply

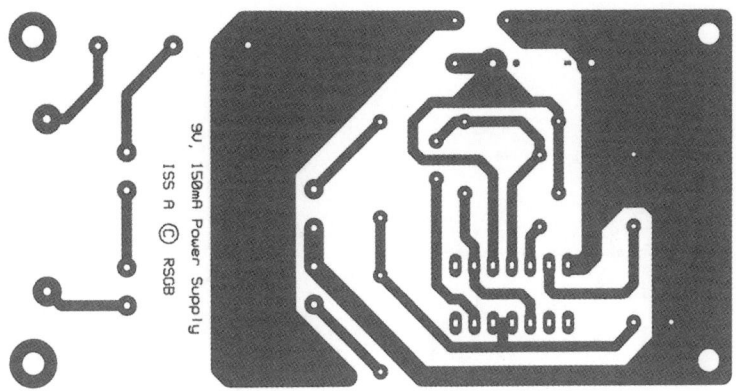

Pattern Fig D9.1

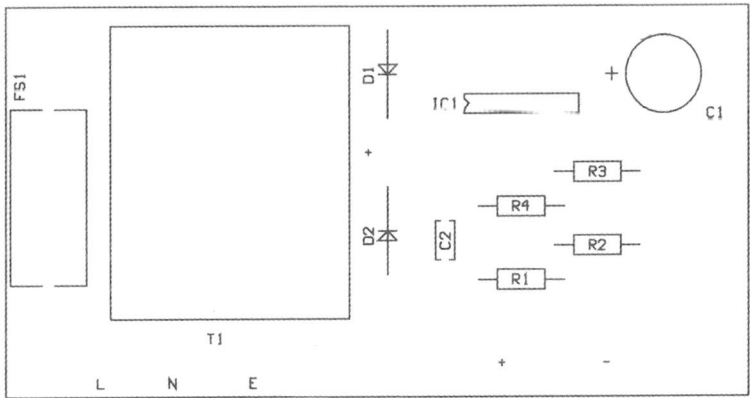

Layout Fig D9.2

Appendix A5.2: Very Low Current Adapter for DC Measurements

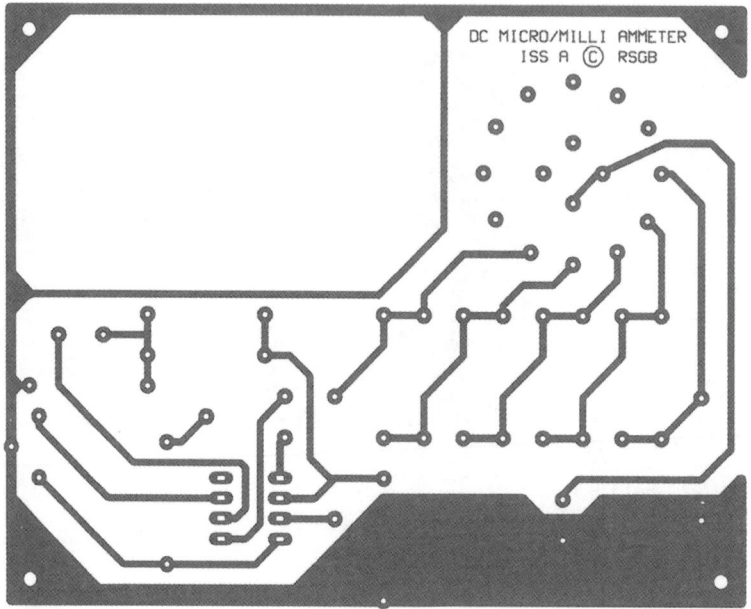

Pattern Fig D11.1

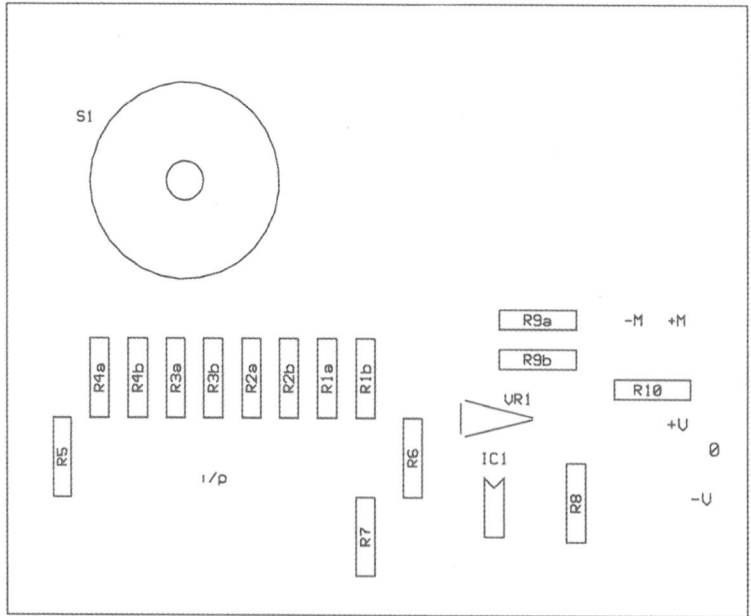

Layout Fig D11.2

Appendix A5.3: A High Impedance AC Voltmeter

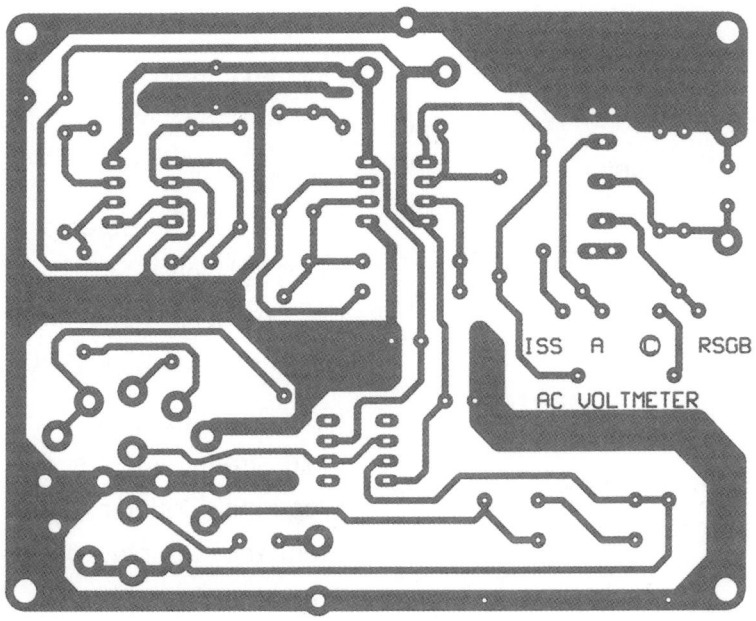

Pattern Fig D12.1

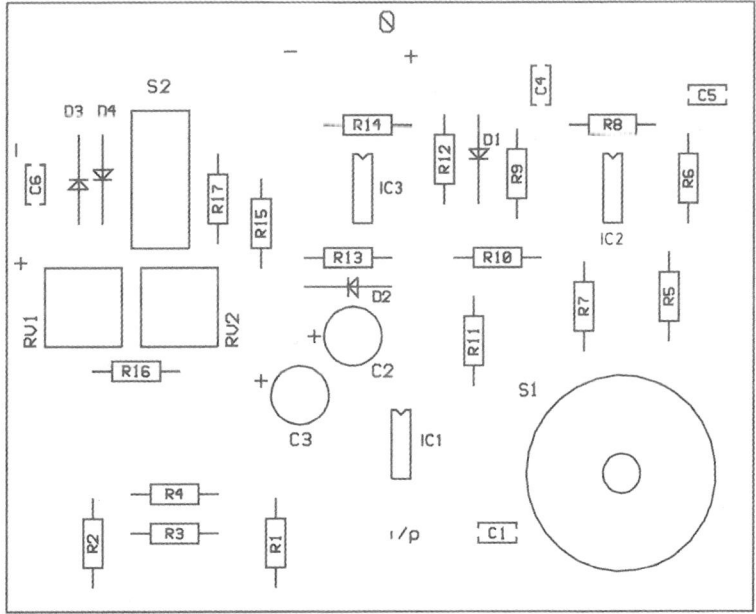

Layout Fig D12.2

Index

Always The Best
Amateur Radio Books

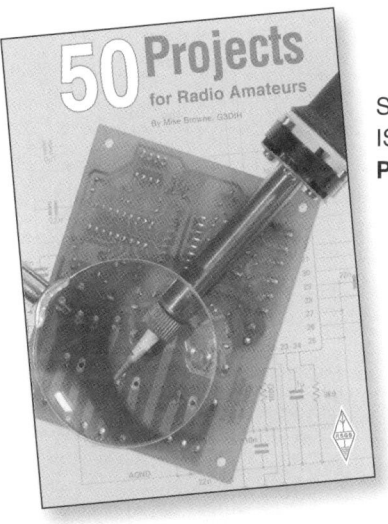

Size 174x240mm, 256pages
ISBN: 9781 9101 9352 5
Price £14.99

Size 174x240mm, 288 pages
ISBN: 9781 9101 9355 6
Price £14.99

Size 174x240mm, 368 pages
ISBN: 9781 9101 9354 9
Price £14.99

Size 174x240mm, 304 pages
ISBN: 9781 9101 9349 5
£12.99

ALWAYS THE BEST
AMATEUR RADIO BOOKS

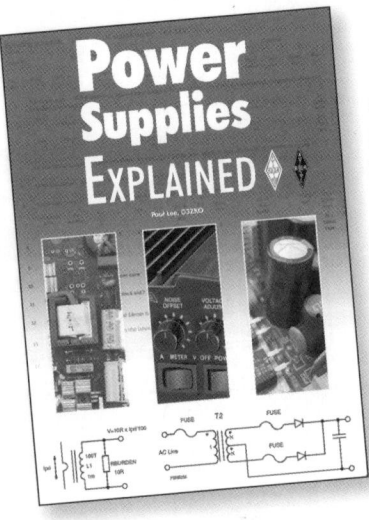